# Water Quality in Drinking Water Distribution Systems

# Water Quality in Drinking Water Distribution Systems

Special Issue Editors

**Mirjam Blokker**
**Joost Van Summeren**
**Vanessa Speight**

MDPI • Basel • Beijing • Wuhan • Barcelona • Belgrade • Manchester • Tokyo • Cluj • Tianjin

*Special Issue Editors*
Mirjam Blokker
KWR Water Research Institute
The Netherlands

Joost Van Summeren
KWR Water Research Institute
The Netherlands

Vanessa Speight
University of Sheffield
UK

*Editorial Office*
MDPI
St. Alban-Anlage 66
4052 Basel, Switzerland

This is a reprint of articles from the Special Issue published online in the open access journal *Water* (ISSN 2073-4441) (available at: https://www.mdpi.com/journal/water/special_issues/Drinking_Water_Distribution).

For citation purposes, cite each article independently as indicated on the article page online and as indicated below:

LastName, A.A.; LastName, B.B.; LastName, C.C. Article Title. *Journal Name* **Year**, *Article Number*, Page Range.

**ISBN 978-3-03936-012-3 (Hbk)**
**ISBN 978-3-03936-013-0 (PDF)**

# Contents

# About the Special Issue Editors

**Mirjam Blokker** is a principal scientist on the drinking water infrastructure team. She is an expert in drinking water demand and developed the SIMDEUM model, which can be used to predict the demand for shower water, toilet flushing water, water consumption, etc. With this model, she also researched the impact of flow speeds and residence times on water quality in the pipeline network and completed her Ph.D. on that topic in 2010. As an extension to this, Mirjam has conducted research and implementation in designing water networks. In recent years, Mirjam researched temperature in pipeline networks and microbiological regrowth in networks. In collaboration with her colleagues on the microbiology team, Mirjam developed models for quantitative microbiological risk analysis (QMRA) for collection, purification, and the distribution network. Mirjam's knowledge of statistics has been applied to pipeline, valve, and fire hydrant failure. Mirjam also stood at the forefront of the introduction of the performance indicator for inadequate supply minutes (OLM). Mirjam owns a few Watershare tools.

**Joost Van Summeren** is a scientific researcher on the water infrastructure team. His research focuses on providing improved insights into the quality and quantity of drinking water in distribution networks and real-time monitoring using sensor networks. His work includes the development and application of new measurement methods and data analysis techniques and numerical water quality modelling. He works on practical research with testing grounds, pilot plants, and living laboratories. For example, he contributed to the development of a drinking water network on a laboratory scale for research on smarter distribution networks. Joost used big data techniques to research the link between brown water reports and temperature, pipeline properties, and demographic factors. An example of applying new measuring methods is identifying substance concentrations in drinking water by interpreting sensor signals using an artificial neural network algorithm. With a background in theoretical geophysics, Joost has knowledge of liquid dynamic processes and numerical modelling techniques. At KWR, he developed his expertise in projects in collaboration with the sector and by participating in international conferences.

**Vanessa Speight**'s primary area of research is water quality transformations in drinking water distribution systems, with a focus on applications of models for decision support, public health risk, and implications for regulatory development. In particular, her interest is understanding the inter-relationships between hydraulics within pipe networks and water quality reactions involving disinfectant residual, disinfection by-product formation, biofilms, indicator organisms, and corrosion. Additional areas of research include the involvement of stakeholders in water policy and management decisions along with water- cycle approaches to sustainability, such as life cycle energy analysis (whole life carbon costing) of water distribution system operation and options for water reuse in urban environments. Prof. Speight is the Managing Director of TWENTY65, an EPSRC-funded consortium with 6 universities and more than 70 industrial collaborative partners working across the water cycle to develop flexible and synergistic solutions tailored to meeting changing societal needs and achieve positive impacts on health, environment, economy, and society.

# Preface to "Water Quality in Drinking Water Distribution Systems"

Safe drinking water is paramount for the health and wellbeing of all human populations. Water is extracted from surface and groundwater sources and treated to comply with drinking water standards. The water is then circulated through the drinking water distribution system (DWDS). Within the DWDS, water quality can deteriorate due to microbiological growth, chemical reactions, interactions with ageing and deteriorating infrastructure, and through maintenance and repair activities. Some DWDS actions may serve to improve water quality; however, these can adversely impact the drinking water system and cause instances of poor water quality or disease outbreaks. We invited papers covering examinations of DWDS design and operational practices and their impact on water quality. We received papers based on practical research in real DWDS and laboratory test facilities. We also received papers on novel modelling approaches. A wide range of water quality aspects was gathered, including temperature, disinfection, bacterial communities and biofilm, (fecal) contamination and QMRA, and the effects of flushing and intermittent supply.

**Mirjam Blokker, Joost Van Summeren, Vanessa Speight**
*Special Issue Editors*

*Review*

# Drinking Water Temperature around the Globe: Understanding, Policies, Challenges and Opportunities

Claudia Agudelo-Vera [1,*], Stefania Avvedimento [2], Joby Boxall [3], Enrico Creaco [2], Henk de Kater [4], Armando Di Nardo [5], Aleksandar Djukic [6], Isabel Douterelo [3], Katherine E. Fish [7], Pedro L. Iglesias Rey [8], Nenad Jacimovic [6], Heinz E. Jacobs [9], Zoran Kapelan [10,11], Javier Martinez Solano [8], Carolina Montoya Pachongo [12], Olivier Piller [13], Claudia Quintiliani [1], Jan Ručka [14], Ladislav Tuhovčák [14] and Mirjam Blokker [1,10]

[1]  KWR-Water Research Institute, 3430 PE Nieuwegein, The Netherlands;
    Claudia.quintilliani@kwrwater.nl (C.Q.); Mirjam.Blokker@kwrwater.nl (M.B.)
[2]  Dipartimento di Ingegneria Civile e Architettura, Università degli Studi di Pavia, 27100 Pavia, Italy;
    stefania.avvedimento01@universitadipavia.it (S.A.); creaco@unipv.it (E.C.)
[3]  Department of Civil and Structural Engineering, University of Sheffield, Sheffield S1 3JD, UK;
    j.b.boxall@sheffield.ac.uk (J.B.); i.douterelo@sheffield.ac.uk (I.D.)
[4]  EVIDES Water utility, 3006 HC Rotterdam, The Netherlands; h.dekater@evides.nl
[5]  Department of Engineering, University of Campania Luigi Vanvitelli–, 81031 Aversa (CE), Italy;
    armando.dinardo@unicampania.it
[6]  Faculty of Civil Engineering, University of Belgrade, 11000 Belgrade, Serbia; djukic@grf.bg.ac.rs (A.Dj.);
    njacimovic@grf.bg.ac.rs (N.J.)
[7]  Sheffield Water Centre, Department of Civil and Structural Engineering, University of Sheffield,
    Sheffield S1 3JD, UK; k.fish@sheffield.ac.uk
[8]  Department of Hydraulic Engineering and Environment, Universitat Politècnica de València,
    46022 Valencia, Spain; piglesia@upv.es (P.L.I.R.); jmsolano@upv.es (J.M.S.)
[9]  Department of Civil Engineering, Stellenbosch University, Stellenbosch 7600, South Africa;
    hejacobs@sun.ac.za
[10] Department of Water Management, Delft University of Technology, 2628 CN Delft, The Netherlands;
    z.kapelan@tudelft.nl
[11] Centre for Water Systems, University of Exeter, Exeter EX4 4QF, UK
[12] Cognita Links, Cali 760033, Colombia; carolina.montoya.pachongo@gmail.com
[13] INRAE, ETBX Research Unit, 75338 Paris, France; olivier.piller@inrae.fr
[14] Institute of Municipal Water Management, Faculty of Civil Engineering, Brno University of Technology,
    612 00 Brno, Czech Republic; Jan.Rucka@vut.cz (J.R.); tuhovcak.l@fce.vutbr.cz (L.T.)
*   Correspondence: Claudia.agudelo-vera@kwrwater.nl

Received: 28 February 2020; Accepted: 1 April 2020; Published: 7 April 2020

**Abstract:** Water temperature is often monitored at water sources and treatment works; however, there is limited monitoring of the water temperature in the drinking water distribution system (DWDS), despite a known impact on physical, chemical and microbial reactions which impact water quality. A key parameter influencing drinking water temperature is soil temperature, which is influenced by the urban heat island effects. This paper provides critique and comprehensive summary of the current knowledge, policies and challenges regarding drinking water temperature research and presents the findings from a survey of international stakeholders. Knowledge gaps as well as challenges and opportunities for monitoring and research are identified. The conclusion of the study is that temperature in the DWDS is an emerging concern in various countries regardless of the water source and treatment, climate conditions, or network characteristics such as topology, pipe material or diameter. More research is needed, especially to determine (i) the effect of higher temperatures, (ii) a legislative limit on temperature and (iii) measures to comply with this limit.

**Keywords:** tap water temperature; climate change; underground hotspots; subsurface urban heat island; water quality and safety; shallow underground

---

## 1. Introduction

A drinking water distribution system (DWDS) is an integral part of a water supply network comprising pipelines, storage facilities and associated assets to carry potable water from treatment plant(s) to water consumers in order to satisfy residential, commercial, industrial and firefighting requirements. One of the most difficult, yet critical, roles of DWDS operation is maintaining microbiological safety for the protection of public health. To guarantee good standards of water quality supply at the end point of the DWDS, many countries maintain a disinfection residual (commonly chlorine) within treated water during distribution. However, several countries in Europe (e.g., The Netherlands [1], parts of Germany [2], Switzerland and Austria [3]) do not use disinfectant residual in the DWDS. They instead rely on catchment protection and highly treated water which includes disinfection via UV light before water enters the DWDS and good DWDS design, operational and maintenance practices. Whether or not a disinfectant residual is present, a variety of water quality reactions are taking place between microorganisms (present in biofilms, sediments and free-floating in the water column), inorganic contaminants, such as corrosion byproducts, and nutrients. These complex reactions are influenced by source water quality (after treatment), hydraulic conditions in the DWDS (driven by customer demands), nature and condition of the infrastructure and temperature [4].

Water quality and hydraulics in the DWDS have been extensively studied [5–9]. Although little is known in practice, research has been conducted to model temperature changes in the DWDS and to determine delivered water temperature at the customer [10–14]. Temperature is an important determinant of water quality, since it influences physical, chemical and biological processes, such as absorption of chemicals, chlorine decay [15] and microbial growth and competition processes [8]. Specifically, it influences the survival and growth conditions of microorganisms and the kinetics of many chemical reactions. Temperature can influence the dynamics of microorganisms in the DWDS promoting the role of biofilms as a reservoir of opportunistic pathogens and their release into the bulk drinking water [16]. Many water treatment processes (e.g., clariflocculation, filtration, ozonation) are influenced by water temperature. However, the applied hydraulic and quality models in the literature usually consider a constant temperature [17,18]. Machell and Boxall [19] highlight the complex interaction of hydraulics (specifically water age), infrastructure conditions and water quality. They specifically show the heating effect of water during its transit through the DWDS during summer months in the UK and the route-specific nature of this. Blokker et al. [20] also analysed this complex interaction when studying the potential to extract thermal energy from drinking water.

Drinking water temperature can significantly increase or decrease during distribution from the source to the customer. This change is strongly influenced by the weather, the depth of installation of transport and distribution pipes, the soil type, ground water levels, presence of anthropogenic heat sources and hydraulic residence times [11,21]. At the building level, drinking water temperature can also be affected by the layout of the hot water installations [14].

The Netherlands is one of the few countries with a specific regulation regarding temperature: the Drinking Water Standards [22] states that the temperature of drinking water at the customers' tap should not exceed 25 °C. Within the regular tap sampling program of the Dutch utilities, in the relatively warm year of 2006, it was reported that 0.1% of samples exceeded the 25 °C limit. With global warming and increasing urbanisation, it is expected that the quantity of samples that exceed the temperature limit will increase.

Over the last decade, Dutch drinking water companies have been researching the impact of drinking water temperature in their DWDSs to guarantee high drinking water quality and to prepare the infrastructure for the challenges that climate change may pose. Despite its importance, according to our best knowledge, only a few researchers [11,23] have developed and published a validated model about how the drinking water temperature changes in the distribution network. In The Netherlands, it was shown that the water temperature at the tap approaches the temperature of the soil that surrounds the distribution mains (pipes with a diameter of 60–200 mm, typical residence times of 48 h or more and located at a depth of 1 m) [11]. In the urban environment, temperatures easily approach the 25 °C limit during a warmer than average summer. Locally, under the influence of anthropogenic heat sources such as district heating pipes or electric cables, the temperature in the DWDS can temporally and locally be higher than 25 °C [21]. Yet, there remains a paucity of information regarding drinking water temperature in the DWDS, especially in countries where temperature limits are not enforced.

This article presents a review of the current knowledge about drinking water temperature from source to tap, as well as a comparison between the policies and practices in a number of countries. Challenges for drinking water companies and policy makers are formulated, resulting in identification of future research directions.

## 2. Methodology

Two methods were used to gather data. A survey was performed to identify local experiences, issues and current knowledge. A literature review was conducted to determine the current scientific knowledge about the potential impact of water temperature on the DWDS.

### 2.1. Survey

A questionnaire was sent to 18 participants of the European Project WatQual (www.sheffield.ac.uk/civil/wat-qual) in August 2018. Participants were researchers from universities or employees of water utilities. The questionnaire contained twelve open questions regarding legislation, practices, knowledge and data about drinking water temperature in the DWDS. Eleven completed questionnaires were returned with data from nine countries. Some answers about practices were anonymised and scientific references were searched to support them. Data from the countries was used to illustrate current practices regarding monitoring.

### 2.2. Literature Review

The online database SCOPUS was used. The search was conducted in November 2019 and no time limit was used in the search. The search was limited to the subject areas: engineering and environmental science. Two searches were conducted. First, the key word "Tap water temperature" was used. This did not provide any relevant documents. A second search using "Drinking water" AND "distribution systems" AND "temperature", focusing on publications in English was performed and 239 articles were found. After a first screening of the articles, only 10 references were relevant for this review. After that, the snowball method was followed, using these key documents as a starting point to find other relevant titles on the subject matter. Additionally, a question regarding relevant literature was included in the questionnaire of the survey. In total, 48 articles from 25 different journals were used in the current study.

## 3. Results—Drinking Water Temperature from Source to Tap

### 3.1. Monitoring Practices

Most of the surveyed water companies systematically monitor source water temperature, and/or the temperature of the treated water (Table 1), as an operational parameter. However, the temperature from source to tap is not systematically monitored in most of the surveyed countries. In the countries where the temperature is monitored, the results are often from discrete samples, resulting in data as shown in Table 1.

**Table 1.** Overview of recorded temperatures SW = surface water, GW = ground water, MW = Mix of GW and SW, RDT: Random Day Time.

| Country | Source | Water Treatment Plant | At the Customer |
|---|---|---|---|
| Colombia [a] | 13–28 °C | 16–26 °C | 25–28 °C |
| Czech Republic | GW: 6–15 °C [b] | SW: 4–11 °C [c] | MW: 2–24 °C [d] |
| France | | GW: 12 °C [e] | RDT: 10 <25 °C [f] |
| Italy | | 6–15°C [g] | |
| Netherlands [h] | | SW: 2–23 °C<br>GW: 12–13°C | RDT: 4–25 °C |
| Serbia | | 9–16 °C [i]<br>SW: 6–27 °C [j]<br>GW: 12–18 °C [j] | 5–18 °C [i] |
| South Africa | | 10–28 °C [k] | 20.5–24.5 °C [l] |
| Spain | | 10–29 °C [m] | |
| United Kingdom | SW: 1–21°C [n] | SW: 2–26 °C [o]<br>GW: 10–18 °C [o]<br>MW: 2–23 °C [o]<br>SW: 3–24 °C [p]<br>GW: 11–12 °C [p]<br>MW: 6–22 °C [p] | SW: 3–25 °C [o]<br>GW: 4–27 °C [o]<br>MW: 4–26 °C [o] |

[a] City of Cali—At the source and water treatment plant: daily measurements, years: 2017–2018, at the tap: nine water samples collected in different days [24]. [b] City of Vsetín, Czech Republic—ground water source, bank infiltration from Bečva river, year 2018–2019. [c] City of Vsetín, Czech Republic—WTP from valley reservoir Karolinka, years 2018–2019. [d] City of Vsetín, Czech Republic—costumer's tap in the city center, years 2018–2019. [e] At Strasbourg—ground water. For other locations, it can exceed 25 °C in some situations. [f] ARS 2020 http://www.eaupotable.sante.gouv.fr. Exceedances of the reference temperature (25 °C) on the water of the distribution networks are frequent in the summer period (2017 results: 138 noncompliant values out of 800 samples taken in June, July and August and 3500 during the year; source ARS). [g] Campania, Southern Italy. [h] Rotterdam, tap samples—RDT, years 2008–2012. [i] Measurements in the city of Pancevo, Serbia, between Feb 2017 and Jan 2018 at three locations: two at the city center and the third a village nearly 18 km from the WTP. [j] Belgrade, Serbia—Years 2013–2018. [k] Nonsystematically monitored. [l] Jacobs et al. [25]. [m] City of Murcia, Spain. Year 2009. Measurements in the water treatment plants and in the network. [n] Dŵr Cymru Welsh Water—years 2010–2017. [o] Anglian Water—2018. [p] Bristol Water—2018 daily measurements.

A few countries monitor water temperatures at the tap. From the surveyed countries, Czech Republic, France, The Netherlands, Serbia and the UK monitor the tap water temperature. This monitoring is usually random and a standard thermometer is used. These samples collected at customers' taps are discrete data sets and are very temporally and spatially sparse. Only in The Netherlands and the Czech Republic is temperature measured and recorded to comply with regulatory reporting requirements. In other countries, it is common that temperature is measured when discrete samples are collected at customer taps, for example, as part of chlorine residual measurements, but the values are not typically recorded or reported. In The Netherlands, the reading is made from the closest tap to the water meter (usually in the kitchen sink on the ground floor). The stagnant water in the domestic installation is flushed; after flushing, the temperature stabilises and it is recorded. In the UK, the standard procedure for random day time sampling is to run the tap for one minute prior to sample collection. In France, The Regional Health Agency (French ARS) randomly checks water temperature at consumers' water taps, where number and frequency of measurements depends on the size of the water utility. In the Czech Republic, the analysis at the consumer's tap also includes measuring the water temperature. The results of the analyses are then electronically sent to the common national PiVO database. The IS PiVo system was created in 2004 as a tool of hygienic service for water quality monitoring in the Czech Republic. All operators of public water supply systems are obliged to monitor the quality of drinking water by law. The results are provided electronically and processed statistically on an annual basis [26]. Table 1 shows the range of measured drinking water temperatures in the surveyed countries.

*3.2. Drinking Water Temperature at The Source*

Source water temperatures have a limited impact on the temperature at the tap. Measurements in The Netherlands have shown that temperature at the customer's tap is mostly determined by the temperature of the soil around the distribution mains (typically at 1.0 m depth in The Netherlands and much of the world), independent of the water source type. Figure 1 shows two supply areas with different water sources: one with a ground water (GW) source, one with a surface water (SW) source; the temperature profiles are unique for each source. Water temperatures at the tap for these areas were also analysed and showed similar temperatures with a seasonal pattern between the two different water sources. These results confirm that the water temperature at the tap is to a limited extent determined by the temperature at the source/outlet of water treatment plant (WTP).

Drinking water temperature at the point where source water (after treatment) enters the DWDS is determined by the type of source water (ground water or surface water) and the characteristics of the facilities where the water is treated and treated water is stored. As a general rule, groundwater temperature is mainly stable over the year. For example, groundwater temperature in The Netherlands is around 12–13 °C, but seasonal temperature variations can be higher if the source is close to a river and groundwater is influenced by riverbank filtration (see Table 1, data for ground water in Belgrade, Serbia). Meanwhile the surface water temperature has high seasonal variations, and its typical pattern is shown in Figure 1. Table 1 shows an overview of variations of water temperature after treatment from different sources, recorded in various countries.

*3.3. Drinking Water Temperature in the Transport and Distribution System*

The temperature gradient between soil surrounding the water main and water in the pipe drives temperature change in the DWDS. The temperature of the shallow underground soil (1–2 m depth), where drinking water mains are often installed, shows seasonal variations. The 'frost depth' is the depth to which the ground water in the soil is expected to freeze in subzero conditions, and it depends on climatic conditions. Frost depth is considered in many countries to determine the minimum installation depth of drinking water mains to avoid freezing of water in the pipes, or breaking pipes from freezing and thawing of the soil around the pipes. Typical installation depths in central Europe vary between 0.8 m and 1.5 m, whereas in countries such as Finland at higher latitudes, installation depths increase, up to 2.5 m. In other countries, where frost is not an issue, the minimum depth of the trenches is determined in such a way that the pipes are protected from traffic and external loads. In Cali (Colombia) an installation depth between 1.0 m and 1.5 m was reported. In Spain, for instance, the minimum depth will be such that the upper border of the pipeline is at least one meter from the surface; under sidewalks it should be a minimum of 0.60 m. In South Africa, the cover should be no less than 0.9 m [27], although older South African standards stipulated 0.6 m minimum cover. Pipes in South Africa are typically installed at approximately 1.5 m. Water reticulation design guidelines provided by WaterCare in New Zealand suggest 1.0 m cover in roads and 0.75 m in berms and open country [28].

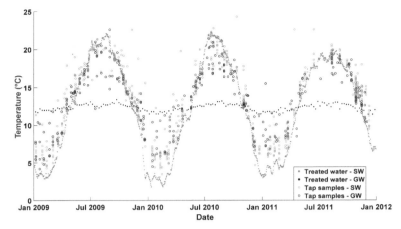

**Figure 1.** Measured water temperature at two pumping stations in The Netherlands—one from surface water (SW) and the other from ground water (GW)—and the respective temperatures at the tap measured at random locations in the separate drinking water distribution system (DWDS) [29].

Soil temperature is influenced by the weather (air temperature, solar radiation, etc.), the environment (rural vs. urban), land-cover (bitumen/tar vs. natural vegetation), soil type and conditions (sand vs. clay and moisture content), as shown below. The energy transfer rate from the soil to the inner pipe wall is determined by the conductivity of the pipe material and the thickness of the pipe wall. Subsequently, the energy is transferred from the inner wall to the flowing water. Within a few hours, drinking water reaches the surrounding soil temperature, depending on factors such as the pipe diameter, wall thickness and flow velocity. Based on the equations presented by Blokker and Pieterse-Quirijns [11] it is possible to calculate the time needed to warm up the water contained in a pipe of a certain diameter, given an initial drinking water temperature and the soil temperature. Figure 2 shows the number of hours needed for drinking water in distribution pipes to heat up from 15 °C to 25 °C and number of minutes in connection pipes to warm up from 20 °C to 25 °C. Plastic and asbestos cement pipes are thermal insulators and this means a relatively long heating time. Cast iron pipes, even with cement lining, show a much shorter time for the water to heat up from 15 to 25 °C for the same diameters, e.g., less than 1 h for a 150 mm cast iron pipe with cement lining [11].

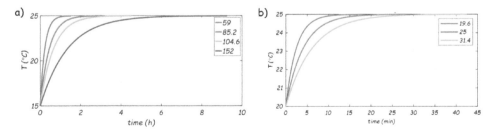

**Figure 2.** Heating up of the drinking water temperature in (**a**) polyvinyl chloride (PVC) distribution pipes with inside diameters between 59 mm and 152 mm. Original water temperature is 15 °C and soil temperature is 25 °C; (**b**) plastic connection pipes with inside diameters between 19.6 mm and 31.4 mm. Original water temperature is 20 °C and soil temperature is 25 °C.

The term "urban heat island" describes built up areas that are hotter than surrounding rural areas due to limited evapotranspiration, heat storage in buildings and urban surfaces, and anthropogenic heat sources. Sources of anthropogenic heat include cooling and heating of buildings, manufacturing, transportation, lighting, etc. [30,31]. Recently it was proven that the temperature of the shallow underground is also strongly influenced by anthropogenic heat sources such as district heating pipes, electricity cables, underground parking garages, etc. and it can lead to which is known as the 'subsurface heat island effect' [32–34]. Analysis of German cities has shown that superposition of various heat sources leads to a significant local warming [32]. Measurements of soil temperatures in The Netherlands have shown that soil temperatures at depth of 1.0 m in a warmer than average summer with a heat wave can reach very local up to 27 °C and can heat up at a rate of 1 °C per day, in so-called 'hot-spot' locations. Examples of 'hot-spot' locations are industrial areas with large anthropogenic heat sources, with no vegetation and good drainage that prevents infiltration and fully exposed to the sun radiation [21].

Blokker et al. [10] modelled drinking water temperature in the DWDS using EPANET-MSX [35]. The use of EPANET-MSX facilitates the calculation of temperature at each node in the distribution network. The model was developed assuming a constant soil temperature over 24 h. Figure 3 shows that tap temperatures vary from 10 °C close to the WTP to 25 °C further downstream. Machell and Boxall [19] reported measured temperatures in the networks and showed that temperature increases with increasing water age along flow routes. Figure 4 shows different pipe routes for a network with two Service Reservoirs (SRs) and demonstrates a range of temperature increases. Although several soil temperature models for rural areas have been proposed, little is known about the soil temperature profile in urban areas. A one-dimensional soil temperature model was developed by Blokker and Pieterse-Quirijns [11] and extended by Agudelo-Vera et al. [21] to include anthropogenic heat sources, as seen in Figure 5.

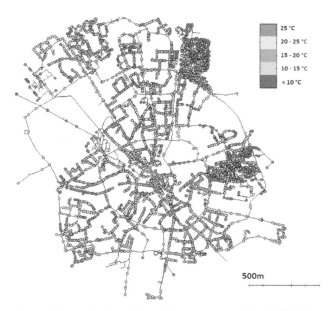

**Figure 3.** Example of simulation of the temperature in the DWDS [10].

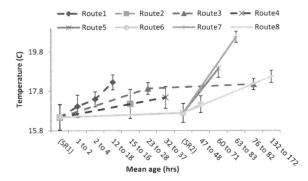

**Figure 4.** Measured temperature versus the calculated water mean age along flow routes for a network with two Service Reservoirs (SRs), with permission from the American Society of Civil Engineers (ASCE) [19].

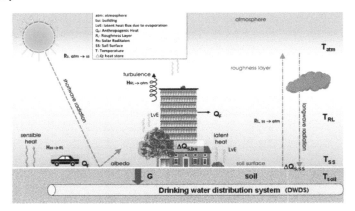

**Figure 5.** Schematic representation of the one-dimensional soil temperature model, with permission from Copernicus Publications [21].

*3.4. Drinking Water Temperature in the Domestic Drinking Water Installation and at the Tap*

Domestic drinking water supply systems (DDWSs) are the final step in the supply of drinking water to consumers. Drinking water temperatures are generally higher in households and buildings than in the distribution system. Drinking water temperature in the domestic drinking water installation can increase due to pipes installed through heated rooms or nearby heat sources [36,37]. Zlatanovic, et al. [14] developed a model to simulate the temperature in DDWSs. The model showed that inlet water temperature and ambient temperature both have a large effect on the water temperature at the household tap.

Drinking water temperature in tropical countries could be even higher; nine water samples collected on different days in the city of Cali (Colombia) resulted in temperatures between 25 °C and 28 °C [24]. Measurements in South Africa showed that the final temperature at the cold water tap varied from one day to the next with a range of ± 6 °C. Spot measurements made in summer with the cold water end-use temperature in one home peaked at 34 °C in an afternoon on December 2018, after a few seconds of the tap running. Temperatures up to 41 °C degrees have been measured during the first 10 s after opening the cold water tap during a very hot midsummer day in Cape Town (January 2020) with an outside air temperature of 42 °C [38]. These relatively high values could be ascribed to the shallow buried plumbing pipe (300 mm ground cover) passing around the Northern side of the house in the full sun [25]. Drinking water temperature without flushing in the DDWS can reach the indoor temperature, in countries where homes are typically not climate controlled, such as South Africa.

*3.5. Drinking Water from Source to Tap*

Water has a relatively large heat capacity; therefore, considerable amounts of energy are required to heat up water. Additionally, water has a relatively high heat transfer coefficient, so it takes some time for the water to heat up; note that the time required to reach a certain temperature is decreased by convection (i.e., flowing water enhances heat transfer). A heat transfer model can calculate that it takes tens of hours to heat up water in a reservoir or a transport main (pipe diameter 300–800 mm), a few hours in a distribution pipe (diameter 60–150 mm), and a few minutes in a property connection pipe (diameter 15–30 mm). This is shown in Figure 2 and [11]. This simple heat transfer model assumes that the driving force is the temperature at the pipe wall, which is not affected by the temperature of the drinking water. This means that the temperature of the pipe wall can be assumed to be equal to the undisturbed soil temperature at installation depth. The undisturbed soil temperature can easily be determined by a one-dimensional micrometeorology model. However, there is a heat exchange between soil and drinking water.

However, as drinking water pipes distribute water of varying temperatures (5–25 °C throughout the year due to seasonal variation), the soil temperature around the drinking water pipe is also affected by the drinking water temperature. As the pipes are installed for a long period of time, it can be expected that the soil temperature around the pipes is not always equal to the undisturbed soil temperature. Thus, the soil temperature around the drinking water pipe is also affected by the drinking water temperature. The interactions between and within the soil temperature and water temperature are complex. The effect of soil temperature on short and long wave radiation, surface convection, and heat transfer through the soil need to be considered in combination with the effect of drinking water temperature, which is difficult to model. The weather-related variables have a seasonal temporal resolution, whereas the drinking water temperature could change in a few hours depending on the flow rate of the water through the pipe.

Given the above and considering the typical residence times of water in the various parts of the network between source and tap, drinking water temperature at different locations between the source and a tap is estimated as follows:

Drinking water temperature at source or treatment plant (Table 1): Temperature is often measured here, and hence known. Ground water temperature at the source will be relatively stable (e.g., 12–13 °C in The Netherlands and U.K./Bristol) year-round, and surface water source temperature can vary substantially between 2 and 27 °C.

Drinking water temperature in the transport main: Typically, almost equal to source/treatment plant temperature (difference of ± 1 °C). Firstly, these mains have a large diameter and are usually short enough for the residence time to be much smaller than the heating time given in Figure 2. Secondly, these large mains substantially influence the surrounding soil temperature, which means there is a limited net heat exchange between the soil and water in the pipe. Furthermore, these mains are typically installed deeper than distribution mains, hence the soil temperature is less affected by the weather.

Drinking water in SRs/tanks: The large volume to surface area of most SRs compared to pipes leads to slower heating/cooling effects during the residence within these critical structures. However, they often have very long residence times. Figure 4 shows the relative impact of flow routes, including a second large SR to retard heating effects during the summer in the UK. It should be noted that this was for an underground tank in a hilly area. Underground tanks are affected by ground temperature as with the buried pipes. Where topology is flatter, such tanks are typically elevated above the ground. In above-ground reservoirs, heating and cooling effects can be very significant due to bigger and more rapid variations in air temperature than in soil temperature. Temperature in the reservoirs can be also affected by material. However, there is not enough data to quantify the level of difference.

Drinking water temperature in the distribution mains: typically quickly approaches the undisturbed soil temperatures at installation depth (typically 1.0 m). These mains have a limited diameter, where the residence time is greater than the heating time from Figure 2. As these mains influence the surrounding soil temperature to a limited extent, the actual heating time may be longer than that shown in Figure 2, but the residence times have the same order of magnitude, so there is significant heat exchange. These mains are typically installed at a depth of 1 m, where the soil temperature is subjected to seasonal change.

Drinking water temperature in the connection water supply pipes: typically almost equal to the temperature at the end of the distribution main (so soil temperature at depth of 1 m). Firstly, these small diameter mains have short lengths, where the residence time (during flow, the situation of stagnant water is kept out of the analysis) is much smaller than the heating time. These small mains hardly influence the surrounding soil temperature, and if they do, the equilibrium would be towards the temperature of the distribution mains. These pipes are typically installed at a shallower depth than distribution mains, so the soil temperature is more influenced by the weather.

Drinking water temperature in the premises plumbing pipes: Typically almost equal to the temperature at the end of the connection and thus of the distribution main (i.e., soil temperature at depth of 1 m). These small diameter mains have short lengths, hence their water residence time (again during flow) is much smaller than the heating time. These mains are not located in the soil, but in airshafts, and the air temperature is not affected by the drinking water temperature of these small-diameter pipes.

Drinking water temperature at the tap: Typically (during flow after flushing) equal to temperature at the end of the distribution mains (i.e., soil temperature at the depth of 1 m) when customers are directly connected to the network. For situations where storage occurs between the distribution network and the customer's tap, other temperatures apply depending on the type of storage (roof or underground), local climate and storage times. Stagnant water will reach the surrounding temperature.

Consequently, it is clear that the soil temperature at the installation depth of the distribution mains is important to know. This temperature is determined by, on one hand, short and long wave radiation (including from above ground anthropogenic sources), surface convection, and heat transfer through the soil and, on the other hand, by the underground (anthropogenic) heat sources. As the anthropogenic sources can have a local effect, it is not easy to predict drinking water temperatures in the entire network. Tap temperatures are not typically measured (Table 1), and in the soil/ground water, only on a project basis. Therefore, the soil temperatures at installation depth are mostly unknown.

*3.6. Consequences of Higher Temperatures and Legislation*

The World Health Organization (WHO) guidelines recommend a maximum temperature limit of 25 °C at the tap [39]: "Cool water is generally more palatable than warm water, and temperature will impact on the acceptability of a number of other inorganic constituents and chemical contaminants that may affect the taste. High water temperature enhances the growth of microorganisms and may increase taste, odour, colour and corrosion problems". In a recent review the WHO reports that in a survey of 104 countries, 18 countries have a regulatory/guideline value of temperature [40]. This review also states that "None of the values for temperature were mandatory, being guiding levels or operational goals. None of the countries and territories' documents indicated what would happen if temperatures rose above the suggested value. In addition to those with numerical values, seven countries and territories had descriptive levels such as: 2.5 °C above normal; "not objectionable"; "air temperature plus 3 °C"; "acceptable"; and "ambient"". No additional information about the countries or the type of standard is given. In the survey conducted for this paper, a number of legal standards were identified, as summarised in Table 2.

**Table 2.** Legal standards and monitoring of the surveyed countries.

| Country | Legal Standard for Drinking Water Temperature | Legal Standard for *Legionella* |
| --- | --- | --- |
| Colombia | No legal standard | No legal standard |
| Czech Republic | Decree No. 252/2004 Coll. Decree laying down hygiene requirements for drinking and hot water and frequency and scope of drinking water control. The recommended temperature of drinking water at the customer's tap is between 8 and 12 °C. | Decree No. 252/2004 Coll. Decree laying down hygiene requirements for drinking and hot water and frequency and scope of drinking water control. This indicator is only set for hot water, where the limit of 100 HTP/100 mL is mandatory. This is the limit that applies to health and accommodation facilities, hot water supplied to showers of artificial or natural pools and drinking water used for hot water production; for other buildings, it is the recommended value to be sought through technical measures. The limit 0 HTP/100 mL as the highest limit value applies to wards of hospitals where immunocompromised patients are located |
| France | The temperature at the consumer's tap should be less than 25 °C (decree from 11 January 2017) in metropolitan France. | For water heating systems of public premises (hospital, hostel, camping, retreat houses, etc.) and cooling towers there is a regulation for environmental monitoring of *Legionella*. Since 1 August 2012, monitoring has been based on culture methods (as per Standard NF T90-431 "Detection and enumeration of *Legionella* spp. and of *Legionella pneumophila* by culture in agar media". However, there are several detection and enumeration methods for *Legionella* that are under development or that are currently in use to greater or lesser extents. Since 1 January 2012, monitoring is mandatory on hot water networks for establishments receiving the public ANSES (French Agency for Food, Environmental and Occupational Health & Safety). |
| Italy | No legal standard. However, it is recommended that temperature should range between 12 °C and 25 °C (Rapporti ISTISAN 97/9, Istituto Superiore della Sanità) | National guidelines from Conferenza Stato-Regioni del 07 maggio 2015. Drinking water temperature must be controlled to be outside of the critical range 20–50 °C to prevent *Legionella* infections. |
| Netherlands | The Dutch Drinking Water Directive contains a maximum temperature limit of 25 °C at the tap [22]. | National guidelines concerning prevention of *Legionella* infections that state the drinking water temperature in a building may not exceed 25 °C, and hot water temperatures must be at least 55 °C [41]. |
| Serbia | Drinking water quality standards (Official gazette of FRYu, No. 42/98 and 44/99, Official gazette of RS No. 28/19), temperature at the consumer is not set, but there is a requirement that it shall not be higher than the temperature at the source. | No standards |

**Table 2.** *Cont.*

| Country | Legal Standard for Drinking Water Temperature | Legal Standard for *Legionella* |
|---|---|---|
| Spain | No standards | There are two laws that establish some parameters related to *Legionella*: (a) Royal Decree 140/2003 of February 7th [42] establishing the sanitary criteria for quality of water for human consumption. In this law, there is no mention to temperature nor *Legionella* at all, but fixes all the values applied to suitable drinking water. It also fixes that sampling protocols for every water company. (b) Royal Decree 865/2003, of 4 July [43] establishes hygienic-sanitary criteria for the prevention and control of Legionnaires' disease. The aim of this law consists of preventing and controlling legionellosis by adopting hygienic and sanitary measures in those facilities where *Legionella* can proliferate and spread. In this sense, it focuses on hot water facilities inside the buildings. The Building Technical Standards (CTE from its initials in Spanish) for the design of plumbing installations inside buildings CTE-DB H4 are based on the aforementioned law. There is a nonmandatory recommendation for drinking water to be under 20 °C where weather conditions allow. |
| South Africa | No standards | No standards around the presence of *Legionella* in drinking water. The National Institute for Communicable Diseases [44] recommends: "The proper design, maintenance and temperature of potable water systems are the most important method for preventing the amplification of *Legionella*. Hot water should be stored above 60 °C and delivered to taps above 50 °C. Cold water should be stored below 20 °C, and dead legs or low flow areas eliminated." Legionnaires disease is a notifiable health condition (compulsory notification) in South Africa. |
| United Kingdom | No standards. The Water Fittings Regulations Guidance book advises to try and keep water supplied to 20 °C as a maximum. | Health and Safety England (HSE) have produced a document which is an "Approved Code of Practice" regarding controlling *Legionella* in water systems. The risk assessment, prevention and control of *Legionella* falls under the 1974 Health and Safety at Work Act (HSWA) and a framework for this assessment is covered by the Control of Substances Hazardous to Health Regulations 2002 (COSHH) [45]. Guidelines suggest control measures of:<br><br>• Cold water stored <20 °C and distributed to all outlets at <20 °C within two minutes of operation<br>• Hot water stored at 60 °C and distributed to outlets at >50 °C within 1 min of operation |

Factors such as nutrient concentration, temperature and pH determine microbial community structure and potential for regrowth within DWDSs. Consequently, changes in temperature in DWDSs can influence microbial community composition, promoting the presence of pathogens and the potential for microbial regrowth, particularly of biofilms in the pipe environment [29,46]. A temperature increase of drinking water can influence the microbial ecology of DWDSs, affecting parameters such as potential growth (e.g., colony count at 22 °C, bacteria of the *coli* group and *Legionella*) and the presence of undesirable microorganisms because of their possible role in disease [29]. There is a difference in the effect of temperature on microorganisms depending on location, either as free-living planktonic organisms in the bulk-water, or as a community within a biofilm attached to the pipe wall. The effect of temperature may also depend on water quality (e.g., disinfectant residual, organic loading) and hydraulics. For example, some microorganisms have their optimal growth at 20 °C, others at 25 °C, and yet others at 30 °C. Thus, the temperature will affect the composition of the biofilm. However, publications about microorganisms in water supplies in many cases do not provide accurate data on water temperature [47]. It has been shown in a chlorinated DWDS in the UK that a rise of temperature from the average 16 °C in the warmer months to a temperature of 24 °C promoted changes and loss in the complexity of microbial biofilm communities [46].

The main concern regarding the impact of temperature increases in DWDS is the potential for the proliferation of pathogens such as *Legionella* spp. Legionellosis is a collection of infections that emerged in the second half of the 20th century, and that are caused by *Legionella pneumophila* and related species of bacteria belonging to the genus *Legionella*. Water is the major natural reservoir for Legionellae, and these bacteria are found worldwide in many different natural and artificial aquatic environments, such as cooling towers, water systems in hotels, domestic water heating systems [48], ships and factories, respiratory therapy equipment, fountains, misting devices, and spa pools [49]. Whether or not disinfectant is used, controlling *Legionella* spp. in a drinking water installation can be problematic [50]. Temperature control is a known measure to prevent the proliferation of *Legionella*. The WHO states that to prevent *Legionella* infection, the recommended temperature for storage and distribution of cold water is below 25 °C, and ideally below 20 °C. Table 2 shows that this recommendation has not been adopted everywhere. Table 2 also shows that temperature standards of building owners are not always matched with temperature standards for drinking water utilities. Laboratory studies of mutant *Legionella* strains show that the bacteria may grow below 20 °C under certain conditions [51]. *Legionella* will survive for long periods at low temperatures and then proliferate when the temperature increases, if other conditions allow.

When temperatures remain below 25 °C, it is expected that growth of *Legionella pneumophila* will not occur or will be limited, whereas at temperatures above 30 °C, it is likely that growth of *Legionella pneumophila* will occur at significant levels, providing the biofilm concentration in the drinking water distribution system is high enough. Another prerequisite for the significant growth of *Legionella pneumophila*, is that the temperature has to be higher than 30 °C for a prolonged period, reported as more than seven days [29].

The results of the survey conducted herein showed that seasonal increase of temperatures can cause unpleasant taste on the palate, which may be related to pipe material (e.g., black alkathane pipework, or lead plumbing pipes). Drinking water companies are generally aware that potential issues can include the occurrence of infections (such as *Salmonella*, *Legionella*, Mycobacterium), chlorine decay and formation of byproducts. As expressed in one survey response " ... it is known that increased water temperature leads to increased biofilm activity in distribution network". Research in The Netherlands on the influence of temperature on discolouration risk, concludes that it is likely that higher temperatures in the DWDS can augment discolouration risk [52,53]. In a tropical DWDS in the city of Cali (Colombia), the formation of disinfection byproducts was clearly influenced by pH, temperature, chlorine dosage, and water age. The interactions observed between these parameters and Trihalomethanes (THMs), were also shaping the microbial characteristics of these systems [24]. Other studies regarding the effects of temperature in the DWDS are reported in Table 3.

**Table 3.** Scientific studies on the effects of temperature in the distribution network or at the tap.

| Aspect | Location | Reference |
|---|---|---|
| Changes in bacterial dynamics | Network | [19,54–56] |
| Increased chlorine decay | Network | [15,19,57] |
| Increased discolouration risk | Network/tap | [19,52,53,58] |
| *L. pneumophila* and opportunistic pathogens | | [59] |
| Seasonal shifts in bacterial communities | Effluents of treatment utilities | [60] |
| Trihalomethanes propagation in DWDS | Network | [61] |

## 4. Challenges and Opportunities

*4.1. Trends*

Increasing urbanisation and climate change seem to be the most important current trends affecting drinking water temperature. The 'urban heat island' has been an object of studies during the last decades, but only recently was it shown that it also affects the shallow subsurface, where DWDS pipes are located. In an urban environment with numerous anthropogenic heat sources, the ground is warmer than it is in a rural area. This also influences the temperature of drinking water and therefore the water quality. Although the biggest impacts of climate change will be felt many years from now, it is important to consider the long life-span of a water distribution network and the potential impacts on infrastructure integrity and water quality management. The replacement of water mains offers the opportunity to improve the network by, among other things, starting to take the impact of climate change into account now. For places where replacement is not feasible, and considering that climate change and water shortages are likely to influence the way water is used and stored, it is important to understand the potential consequences of elevated temperature to manage their risks in alternative ways.

Currently, during hot summers, there are concerns when water temperatures exceed 20 °C due to the increased risk of *Legionella* proliferating in premises water systems. With climate change and urbanisation, it is expected that drinking water temperatures will rise [29,62]. As there is hardly any monitoring being done, it is not easy to actually see this trend occurring. The effect of higher water temperatures (on health, organoleptic parameters) is not known. In some countries, this means that legislation is on the "safe side" and limits the drinking water temperature to 25 °C. However, it is not easy to guarantee water supply below this temperature. Firstly, there is no monitoring programme, so compliance is largely unknown and hard to enforce. Secondly, when there is a noncompliance, there is no easy operational measure available to resolve the issue. Flushing can work locally, but at the network scale, the system may not have enough pressure capacity to drastically shorten the residence times [63] and it provides only short or very short-term amelioration. Forensics to quickly determine where the problem is introduced upstream do not exist and when the problem location is determined it may be expensive to solve, too late to react and difficult to determine where the liability lies. Thirdly, when there is a large noncompliance, i.e., the problem is not local but instead occurring in the whole network, there is no operational measure available at all to resolve the issue. The only solution would be to install pipes deeper or take other (large scale) design and installation measures to ensure less effect of climate change or urbanisation on high soil temperatures. Alternatively, we could accept the inadequacy of DWDS and, for example, advocate solutions such as point-of-treatment via small packaged UV systems. Such systems are commercially available and, anecdotally, increasingly common in countries such as South Korea. However, the social, moral, and regulatory implications of such an approach are dramatic and far-reaching.

Another factor to consider is the increasing use of smart appliances and other water saving/demand management type technologies (e.g., rainwater harvesting, grey water recycling, smart meters, etc.). These technologies are likely to affect water temperature at different locations in the system, from property level to pipe network level. For example, increased use of rain or grey water may reduce potable water demand, increasing domestic plumbing and DWDS residence times, and increasing summer months' heating effects from the surrounding air and ground, respectively. The impact of these technologies on water temperature is not currently well understood. Greywater poses an increased risk as it originates from heated sources in the home such as the shower, bath or clothes washing machine, with a notably increased temperature of the reused greywater, often combined with relatively poor quality when compared to water from the DWDS [64]. Alternatively, smart appliances may be managed to use water at specific times and locations to limit residence times by managing the flow through DWDS and premises to avoid peak high temperatures.

Other future changes in the urban environment (e.g., wider use of geothermal energy, district heating systems, etc.) and related planning which is increasingly done in an integrated way, based on the principles of circular economy and water–energy nexus type thinking, may result in further alterations of water temperatures in the built environment and consequentially DWDS water quality as well. The impact of temperature and its link to these issues is not understood well.

*4.2. Knowledge Gaps and Future Research*

Further studies on the influence of temperature on the interacting factors impacting drinking water safety as it travels through DWDS infrastructure are needed. Critical amongst these are biofilm structure, potential for disinfection, byproduct formation and overall biological stability. It is important that the effects of temperature are studied as an integral part of the complex physical, chemical and biological processes interacting within DWDSs. This must cover the full range of basic drinking water quality and representative physical conditions: water age, biological stability, Natural Organic Matter (NOM)/Assimilable Organic Carbon (AOC)/Total organic carbon (TOC) content, chlorine, chemical composition, infrastructure materials, surface area to volumes, hydraulic regimes, etc. Typical temperature ranges are usually limited in a specific supply area. Future studies and data gathered from DWDS in, for example, South European countries or in tropical networks such as the one mentioned here from Colombia, can aid to elucidate the global impact of climate change in DWDS. The practical first step could be to determine the effect (health, organoleptic parameters) of higher temperatures, and from this to determine a proper limit on temperature. Ideally, this should be done in an international context with various water qualities and temperatures and considering the local characteristics of the drinking water distribution systems, such as the water quality after treatment, use of chlorination, roof tanks or intermittent distribution.

The next step would be to look at which measures are possible and sufficient to ensure this realistic temperature limit, first using a model-driven approach, and then when sufficient data is available, by a data-driven approach. Things to consider in design and installation, for example, are the minimum distance between district heating pipes or electricity cables and drinking water pipes, the effect of installation depth, soil coverage by grass or shade. Additionally, tools have to be developed to predict and forecast the short-term consequences of heat waves [65,66] or long-term climate change. In this paper, the challenges and threats of higher drinking water temperatures were extensively reviewed. This research did not address opportunities, such as reclaiming thermal energy from drinking water. Some research [12,20,67] has been conducted showing there is potential.

## 5. Conclusions

A range of issues related to water temperature in drinking distribution systems and its potential impact on water quality in these systems was addressed in this paper. The methodology adopted is based on a literature review and a stakeholder survey conducted in nine countries.

Based on the information and results obtained, the following observations are made:

Water temperatures are monitored, but this is not done systematically, and data collected varies substantially across different countries. In most cases, water temperature is most frequently monitored at sources and treatment plants. There is limited and sporadic monitoring in the DWDS. This monitoring should be done more systematically for a number of reasons, including improved compliance testing and underpinning future research in this area. In many countries, temperature is already measured, such as part of when measuring for chlorine residuals on site, but is not recorded. Therefore, such data could readily be gained with minimal additional effort.

It is widely acknowledged in the literature and engineering practice of different countries that a link exists between drinking water temperature and quality—lower temperatures are linked to improved quality. However, this link is currently not well understood for a range of potential water quality issues. This includes the significance of the 25 °C threshold, which water utilities in some countries are already asked to comply with.

Water temperature varies as it travels from the water treatment works to a tap, primarily due to exchange with the surrounding ground and ground water. Whilst plausible models could be proposed to simulate the processes involved, these remain unverified at present. There is a need for anthropogenic heat sources, and pipe hydraulics.

A number of future changes in the surrounding environment are likely to impact the water temperature in the DWDS. These include climate change, urbanisation, more integrated urban planning, rainwater use, greywater reuse and wider application of water saving and other technologies. The impact of these changes on the DWDS temperature and consequential water quality is currently not well understood and requires future research.

The temperature in drinking water distribution systems is clearly an emerging concern in many countries around the world (not just in warmer climates) and hence should be studied more closely in the future and supported via suitable research funding programmes.

**Author Contributions:** Conceptualization and methodology, C.A.V. and M.B.; literature review, C.A.V.; Data collection (survey), S.A., E.C., A.D.N., A.D., I.D., K.E.F., P.L.I.R., N.J., H.E.J., J.M.S., C.M.P., O.P., C.Q., J.R., L.T.; writing—original draft preparation, C.A.V. and M.B.; review and editing, all authors; paper review and consolidation with additional text provided in places—Z.K., K.E.F., H.E.J. and J.B. All authors have read and agreed to the published version of the manuscript.

**Funding:** This research was partially funded by the European project WATQUAL https://www.sheffield.ac.uk/civil/wat-qual and partly by the joint research program (BTO) of the Dutch Drinking Water Companies.

**Acknowledgments:** We would like to thank Stewart Husband (University of Sheffield) for providing the pathway for dissemination and access to UK water companies and coordinated and gathering responses; Ceyhun Koseoglu (Bristol Water), Paul Gaskin (Dŵr Cymru Welsh Water) and Claire Moody (Anglian Water) for providing data, Marco Dignum (Waternet) for reading and commenting on the final version of the document.

**Conflicts of Interest:** The authors declare no conflict of interest.

## References

1. Smeets, P.W.M.H.; Medema, G.J.; Van Dijk, J.C. The Dutch secret: How to provide safe drinking water without chlorine in The Netherlands. *Drink. Water Eng. Sci.* **2009**, *2*, 1–14. [CrossRef]
2. Uhl, W.; Schaule, G. Establishment of HPC(R2A) for regrowth control in non-chlorinated distribution systems. *Int. J. Food Microbiol.* **2004**, *92*, 317–325. [CrossRef] [PubMed]
3. Rosario-Ortiz, F.; Rose, J.; Speight, V.; Gunten, U.v.; Schnoor, J. How do you like your tap water? *Science* **2016**, *351*, 912–914. [CrossRef] [PubMed]
4. Douterelo, I.; Sharpe, R.L.; Husband, S.; Fish, K.E.; Boxall, J.B. Understanding microbial ecology to improve management of drinking water distribution systems. *Wires Water* **2019**, *6*, e01325. [CrossRef]
5. Blokker, M.; Agudelo-Vera, C.; Moerman, A.; Van Thienen, P.; Pieterse-Quirijns, I. Review of applications for SIMDEUM, a stochastic drinking water demand model with a small temporal and spatial scale. *Drink. Water Eng. Sci.* **2017**, *10*, 1–12. [CrossRef]
6. Douterelo, I.; Sharpe, R.L.; Boxall, J.B. Influence of hydraulic regimes on bacterial community structure and composition in an experimental drinking water distribution system. *Water Res.* **2013**, *47*, 503–516. [CrossRef]
7. Liu, G.; Verberk, J.Q.J.C.; Van Dijk, J.C. Bacteriology of drinking water distribution systems: An integral and multidimensional review. *Appl. Microbiol. Biotechnol.* **2013**, *97*, 9265–9276. [CrossRef]
8. Prest, E.I.; Hammes, F.; van Loosdrecht, M.C.M.; Vrouwenvelder, J.S. Biological Stability of Drinking Water: Controlling Factors, Methods, and Challenges. *Front. Microbiol.* **2016**, *7*, 45. [CrossRef]
9. Sharpe, R.L.; Biggs, C.A.; Boxall, J.B. Hydraulic conditioning to manage potable water discolouration. In Proceedings of the Institution of Civil Engineers-Water Management; Thomas Telford Ltd.: London, UK, 2019; pp. 3–13.
10. Blokker, E.J.M.; Pieterse-Quirijns, E.J.; Vogelaar, A.J.; Sperber, V. *Bacterial Growth Model in the Drinking Water Distribution System—An Early Warning System*; Prepared 2014.023; KWR: Nieuwegein, The Netherlands, 2012.
11. Blokker, E.J.M.; Pieterse-Quirijns, I. Modeling temperature in the drinking water distribution system. *J. Am. Water Works Assoc.* **2013**, *105*, E19–E28. [CrossRef]
12. De Pasquale, A.M.; Giostri, A.; Romano, M.C.; Chiesa, P.; Demeco, T.; Tani, S. District heating by drinking water heat pump: Modelling and energy analysis of a case study in the city of Milan. *Energy* **2017**, *118*, 246–263. [CrossRef]
13. van der Zwan, S.; Pothof, I.; Dignum, M. Multifunctional design to prevent excessive heating of drinking water. In Proceedings of the IWA WCE 2012, Dublin, Ireland, 13–18 May 2012.
14. Zlatanovic, L.; Moerman, A.; van der Hoek, J.P.; Vreeburg, J.; Blokker, M. Development and validation of a drinking water temperature model in domestic drinking water supply systems. *Urban Water J.* **2017**, *14*, 1031–1037. [CrossRef]

15. Monteiro, L.; Figueiredo, D.; Covas, D.; Menaia, J. Integrating water temperature in chlorine decay modelling: A case study. *Urban Water J.* **2017**, *14*, 1097–1101. [CrossRef]

16. Ingerson-Mahar, M.; Reid, A. *Microbes in Pipes: The Microbiology of the Water Distribution System*; American Academy of Microbiology: Boulder, CO, USA, 2013.

17. DiGiano, F.A.; Zhang, W. Uncertainty Analysis in a Mechanistic Model of Bacterial Regrowth in Distribution Systems. *Environ. Sci. Technol.* **2004**, *38*, 5925–5931. [CrossRef] [PubMed]

18. Fisher, I.; Kastl, G.; Sathasivan, A. A suitable model of combined effects of temperature and initial condition on chlorine bulk decay in water distribution systems. *Water Res.* **2012**, *46*, 3293–3303. [CrossRef]

19. Machell, J.; Boxall, J. Modeling and field work to investigate the relationship between age and quality of tap water. *J. Water Resour. Plan. Manag.* **2014**, *140*, 04014020. [CrossRef]

20. Blokker, E.J.M.; van Osch, A.M.; Hogeveen, R.; Mudde, C. Thermal energy from drinking water and cost benefit analysis for an entire city. *J. Water Clim. Chang.* **2013**, *4*, 11–16. [CrossRef]

21. Agudelo-Vera, C.; Blokker, M.; De Kater, H.; Lafort, R. Identifying (subsurface) anthropogenic heat sources that influence temperature in the drinking water distribution system. *Drink. Water Eng. Sci. Discuss.* **2017**, *10*, 83–91. [CrossRef]

22. Drink Water Directive. Drinkwaterbesluit. Available online: http://wetten.overheid.nl/BWBR0030111/geldigheidsdatum_25-02-2013 (accessed on 2 December 2019).

23. Piller, O.; Tavard, L. Modeling the Transport of Physicochemical Parameters for Water Network Security. *Procedia Eng.* **2014**, *70*, 1344–1352. [CrossRef]

24. Montoya-Pachongo, C.; Douterelo, I.; Noakes, C.; Camargo-Valero, M.A.; Sleigh, A.; Escobar-Rivera, J.-C.; Torres-Lozada, P. Field assessment of bacterial communities and total trihalomethanes: Implications for drinking water networks. *Sci. Total Environ.* **2018**, *616*, 345–354. [CrossRef]

25. Jacobs, H.; Botha, B.; Blokker, M. Household Hot Water Temperature–An Analysis at End-Use Level. Proceedings of WDSA/CCWI Joint Conference, Kingston, ON, Canada, 23–25 July 2018.

26. Novakova, J.; Rucka, J. Undesirable consequences of increased water temperature in drinking water distribution system. *Mm Sci. J.* **2019**, *2019*, 3695–3701. [CrossRef]

27. South African National Standard. South African National Standard. South African National Standard SANS 2001-DP1 Edition 1.1. In *Construction Works Part DP1: Earthworks for Buried Pipelines and Prefabricated Culverts*; 2011 Edition 1.1 South African National Standard—Construction Works Part DP1: Earthworks for Buried Pipelines and Prefabricated Culverts; SABS Standards: Pretoria, South Africa, 2011; ISBN 978-0-626-25160-4.

28. Services, W. *Guidelines for Design of Water Reticulation and Pumping Stations*; Watercare Services Ltd.: Auckland, New Zealand, 2013.

29. Agudelo-Vera, C.M.; Blokker, E.J.M.; van der Wielen, P.W.J.J.; Raterman, B. *Drinking Water Temperature in Future Urban Areas*; BTO 2015.012; KWR: Nieuwegein, The Netherlands, 2015.

30. Herb, W.R.; Janke, B.; Mohseni, O.; Stefan, H.G. Ground surface temperature simulation for different land covers. *J. Hydrol.* **2008**, *356*, 327–343. [CrossRef]

31. Mihalakakou, G. On estimating soil surface temperature profiles. *Energy Build.* **2002**, *34*, 251–259. [CrossRef]

32. Menberg, K.; Bayer, P.; Zosseder, K.; Rumohr, S.; Blum, P. Subsurface urban heat islands in German cities. *Sci. Total Environ.* **2013**, *442*, 123–133. [CrossRef] [PubMed]

33. Menberg, K.; Blum, P.; Schaffitel, A.; Bayer, P. Long-term evolution of anthropogenic heat fluxes into a subsurface urban heat island. *Environ. Sci. Technol.* **2013**, *47*, 9747–9755. [CrossRef]

34. Müller, N.; Kuttler, W.; Barlag, A.-B. Analysis of the subsurface urban heat island in Oberhausen, Germany. *Clim. Res.* **2014**, *58*, 247–256. [CrossRef]

35. Shang, F.; Uber, J.G. *EPANET Multi-Species Extension User's Manual*; EPA/600/S-07/021; EPA: Cincinnati, OH, USA, 2008.

36. Lautenschlager, K.; Boon, N.; Wang, Y.; Egli, T.; Hammes, F. Overnight stagnation of drinking water in household taps induces microbial growth and changes in community composition. *Water Res.* **2010**, *44*, 4868–4877. [CrossRef]

37. Lipphaus, P.; Hammes, F.; Kötzsch, S.; Green, J.; Gillespie, S.; Nocker, A. Microbiological tap water profile of a medium-sized building and effect of water stagnation. *Environ. Technol.* **2014**, *35*, 620–628. [CrossRef]

38. Jacobs, H. Personal Communication. 13 February 2020.

39. WHO. *Guidelines for Drinking-Water Quality*; WHO: Geneva, Switzerland, 2006.

40. WHO. *A Global Overview of National Regulations and Standards for Drinking-Water Quality*; WHO: Geneva, Switzerland, 2018; Handleiding Legionellapreventie in leidingwater. Richtlijnen voor prioritaire installaties.

41. ISSO 55.1. *Handleiding. Legionellapreventie in Leidingwater. Richtlijnen Voor Prioritaire Installaties*; ISSO: Rotterdam, The Netherlands, 2012; pp. 94–95.

42. Real Decreto 140/2003, de 7 de febrero, por el que se establecen los criterios sanitarios de la calidad del agua de consumo humano. In *Boletín Oficial del Estado 2003*; Ministerio de Sanidad y Consumo: Madrid, España, 1980; Volume 45.

43. Real Decreto 865/2003, de 4 de julio, por el que se establecen los criterios higiénico-sanitarios para la prevención y control de la legionelosis. In *Boletín Oficial del Estado 2003*; Ministerio de Sanidad y Consumo: Madrid, España, 2003; Volume 171.

44. Carrim, M.; Cohen, C.; de Gouveia, L.; Essel, V.; Mc Carthy, K.; Stewart, R.; Thomas, T.; von Gottberg, A.; Wolter, N. Legionnaires'disease: Nicd recommendations for diagnosis, management and public health response. *Foreword Contents* **2016**, *14*, 137.

45. Britain, G. *Control of Substances Hazardous to Health Regulations 2002*; The Stationery Office: London, UK, 2002.

46. Preciado, C.C.; Boxall, J.; Soria-Carrasco, V.; Douterelo, I. Effect of temperature increase in bacterial and fungal communities of chlorinated drinking water distribution systems. *Access Microbiol.* **2019**, *1*. [CrossRef]

47. van der Kooij, D.; van der Wielen, P. *Microbial Growth in Drinking Water Supplies. Problems, Causes, Control and Research Needs*; Kooij, D.v.d., Wielen, P.W.v.d., Eds.; Iwa Publishing: London, UK, 2013.

48. Stone, W.; Louw, T.M.; Gakingo, G.K.; Nieuwoudt, M.J.; Booysen, M.J. A potential source of undiagnosed Legionellosis: Legionella growth in domestic water heating systems in South Africa. *Energy Sustain. Dev.* **2019**, *48*, 130–138. [CrossRef]

49. WHO. *Legionella and the Prevention of Legionellosis*; WHO: Geneva, Switzerland, 2007.

50. van der Lugt, W.; Euser, S.M.; Bruin, J.P.; den Boer, J.W.; Yzerman, E.P.F. Wide-scale study of 206 buildings in The Netherlands from 2011 to 2015 to determine the effect of drinking water management plans on the presence of Legionella spp. *Water Res.* **2019**, *161*, 581–589. [CrossRef] [PubMed]

51. Söderberg, M.A.; Rossier, O.; Cianciotto, N.P. The type II protein secretion system of Legionella pneumophila promotes growth at low temperatures. *J. Bacteriol.* **2004**, *186*, 3712–3720. [CrossRef] [PubMed]

52. Blokker, E.J.M.; Schaap, P.G. *Effecten van Temperatuur op Bruinwaterrisico*; KWR 2015.091; KWR: Nieuwegein, The Netherlands, 2015.

53. van Summeren, J.; Raterman, B.; Vonk, E.; Blokker, M.; van Erp, J.; Vries, D. Influence of Temperature, Network Diagnostics, and Demographic Factors on Discoloration-Related Customer Reports. *Procedia Eng.* **2015**, *119*, 416–425. [CrossRef]

54. Francisque, A.; Rodriguez, M.J.; Miranda-Moreno, L.F.; Sadiq, R.; Proulx, F. Modeling of heterotrophic bacteria counts in a water distribution system. *Water Res.* **2009**, *43*, 1075–1087. [CrossRef] [PubMed]

55. Niquette, P.; Servais, P.; Savoir, R. Bacterial Dynamics in the drinking water distribution system of Brussels. *Water Res.* **2001**, *35*, 675–682. [CrossRef]

56. Vital, M.; Dignum, M.; Magic-Knezev, A.; Ross, P.; Rietveld, L.; Hammes, F. Flow cytometry and adenosine tri-phosphate analysis: Alternative possibilities to evaluate major bacteriological changes in drinking water treatment and distribution systems. *Water Res.* **2012**, *46*, 4665–4676. [CrossRef]

57. Li, X.; Gu, D.-M.; Qi, J.-Y.; Ukita, M.; Zhao, H.-B. Modeling of residual chlorine in water distribution system. *J. Environ. Sci.* **2003**, *15*, 136–144.

58. Sunny, I.; Husband, S.; Boxall, J. Seasonal Temperature and Turbidity Behaviour in Trunk Mains. Proceedings of WDSA/CCWI Joint Conference, Kingston, ON, Canada, 23–25 July 2018.

59. van der Wielen, P.W.J.J.; van der Kooij, D. Nontuberculous Mycobacteria, Fungi, and Opportunistic Pathogens in Unchlorinated Drinking Water in The Netherlands. *Appl. Environ. Microbiol.* **2013**, *79*, 825. [CrossRef]

60. Pinto, A.J.; Xi, C.; Raskin, L. Bacterial Community Structure in the Drinking Water Microbiome Is Governed by Filtration Processes. *Environ. Sci. Technol.* **2012**, *46*, 8851–8859. [CrossRef]

61. Li, X.; Zhao, H.-b. Development of a model for predicting trihalomethanes propagation in water distribution systems. *Chemosphere* **2006**, *62*, 1028–1032. [CrossRef]

62. Levin Ronnie, B.; Epstein Paul, R.; Ford Tim, E.; Harrington, W.; Olson, E.; Reichard Eric, G.U.S. drinking water challenges in the twenty-first century. *Environ. Health Perspect.* **2002**, *110*, 43–52. [CrossRef] [PubMed]

63. Blokker, E.J.M.; Pieterse-Quirijns, E.J. *Scenariostudies Voor Beperken Invloed Klimaatveranderingen op Temperatuur en Kwaliteit Drinkwater in Het Net*; KWR 2012.017; KWR Watercycle Research Institute: Nieuwegein, The Netherlands, 2012.

64. Nel, N.; Jacobs, H.E. Investigation into untreated greywater reuse practices by suburban households under the threat of intermittent water supply. *J. Water Sanit. Hyg. Dev.* **2019**, *9*, 627–634. [CrossRef]

65. Agudelo-Vera, C.; Blokker, M.; Pieterse-Quirijns, I. Early Warning Systems to Predict Temperature in the Drinking Water Distribution Network. *Procedia Eng.* **2014**, *70*, 23–30. [CrossRef]

66. Agudelo-Vera, C.M.; Blokker, E.J.M.; Pieterse-Quirijns, E.J. Early warning system to forecast maximum temperature in drinking water distribution systems. *J. Water Supply Res. Technol. AQUA* **2015**, *64*, 496–503. [CrossRef]

67. Hubeck-Graudal, H.; Kirstein, J.; Ommen, T.; Rygaard, M.; Elmegaard, B. Drinking water supply as low-temperature source in the district heating system: A case study for the city of Copenhagen. *Energy* **2020**, *194*, 116773. [CrossRef]

*Article*

# Migration and Transformation of Ofloxacin by Free Chlorine in Water Distribution System

**Weiwei Bi** [1], **Yi Jin** [1] **and Hongyu Wang** [2,*]

[1]  College of Civil Engineering and Architecture, Zhejiang University of Technology, Hangzhou 310023, China;
    weiweibi@zjut.edu.cn (W.B.); jinyizjut@sina.cn (Y.J.)
[2]  College of Environment, Zhejiang University of Technology, Hangzhou 310023, China
*  Correspondence: whyzjut@163.com; Tel.: +86-571-8529-0520

Received: 8 March 2019; Accepted: 16 April 2019; Published: 19 April 2019

**Abstract:** This study investigated the degradation kinetics and product generation of ofloxacin (OFL) in the pipe network under different pipe materials, flow rate, pH, free chlorine concentration and temperature. The experiments done in the beaker and pipe network were compared. The results showed that the reaction rate of OFL chlorination with free chlorine increased with the increase of the free chlorine concentration in the pipe network and deionized water, and the degradation efficiency of OFL in the pipe network was higher than that in the deionized water, satisfying the second-order dynamics model. The degradation rate under different pHs was: neutral > acidic > alkaline. The influence of the flow rate is not significant while the influence of the pipe materials and temperature is obvious. The degradation rate of OFL increased with the increase of the temperature, indicating that the OFL degradation was an endothermic process. A liquid chromatograph-mass spectrometer (LC-MS) was used to detect the chlorination intermediates, and the results showed that the piperazine ring was the main group involved in the chlorination reaction, and the main point involved in the chlorination reaction was the N4 atom on the piperazine ring. We also found that, as the reaction time increases, the concentrations of trihalomethanes (THMs) and haloacetic acids (HAAs) increase and THMs mainly exist in the form of trichloromethane (TCM) while HAAs mainly exist in the form of monochloroacetic acid (MCAA).

**Keywords:** water distribution system; water quality; free chlorine; ofloxacin; degradation; kinetic study; formation mechanism

---

## 1. Introduction

In recent years, antibiotics and fluoroquinolones have been widely used in the treatment of various infectious diseases due to their broad antibacterial spectrum and strong efficacy [1–3]. However, as the use of these antibiotics and fluoroquinolones increases, some of them have been detected in the environment. This will affect the growth and development of the plant, and some of the microorganisms in the environment will have an inhibitory effect and may even be killed [4]. Ofloxacin (OFL) was used as the third generation of the fluoronone and has also been constantly detected in drinking water. Chlorine disinfection is the common method used to reduce OFL while the disinfection by-products, such as trihalomethanes (THMs) and haloacetic acids (HAAs), which have carcinogenic, teratogenic and abrupt effects, have been detected as well [5–9]. There are more than 700 kinds of Chlorination disinfection by-products which have been detected in drinking water and that are harmful to the human body, including more than 20 kinds of carcinogens and 50 kinds of mutants which threatens the drinking water safety and the health of the people [10–16]. Therefore, the migration and transformation rules of OFL and disinfection by-products in the water environment and their harm to the environment have been of wide concern.

While the degradation of OFL has been investigated in previous research, they mainly focused on the simple reaction environment of the beaker deionized water [4,6,7]. The environment in the pipe networks is much more complex, and research of the degradation, migration and transformation process of OFL in the pipe network has not been reported [13–16]. Therefore, this study investigated the degradation kinetics, migration and transformation mechanism of the OFL in the pipe network and the deionized water under the action of the free chlorine to provide a reference for the effective control of OFL in the urban water system and the protection of the human and living body.

## 2. Materials and Methods

### 2.1. Test Materials and Instruments

A number of test reagents have been used in this study, including sodium hypochlorite, sodium hydroxide, sodium thiosulfate, ascorbic acid, phosphate, disodium hydrogen phosphate, sodium dihydrogen phosphate, boric acid, borax, concentrated hydrochloric acid, residual chlorine powder, methanol, acetonitrile, CNWBOND HC-18 SPE (500 mg/3 mL, CNW Technologies GmbH, Düsseldorf, Germany) extraction column, and methyl tert-butyl ether (MTBE). The test instruments that have been employed for the analysis include Agilent-2100 High Performance Liquid Chromatography (HPLC, Agilent Technologies, Santa Clara, CA, USA), an Agilent-6460 liquid phase triple quadrupole mass spectrometer (LC-MS, Agilent Technologies, Santa Clara, CA, USA), a DR 2800 UV spectrophotometer (Shimadzu, Kyoto, Japan), an NW Ultra-cure system Super pure water machine (Sigma Aldrich, Shanghai, China), a PHS-3C type pH meter (Sigma Aldrich, Shanghai, China), a liquid gun (Sigma Aldrich, Shanghai, China), a CNW-12 type 12-hole solid phase extraction device (Sigma Aldrich, Shanghai, China), an ND-200 type nitrogen sweeper (Sigma Aldrich, Shanghai, China), a magnetic electromagnetic agitator (Reagent Co., Ltd. Shanghai, China), an ultrasound (Reagent Co., Ltd. Shanghai, China), a dry box (Reagent Co., Ltd. Shanghai, China), EL204 electronic analysis scales (Reagent Co., Ltd. Shanghai, China), a storage of refrigerators, and a liquid extraction device (Sigma Aldrich, Shanghai, China). All of these experiments were undertaken in a water quality comprehensive simulation platform of a large pipe network, as illustrated in Figure 1.

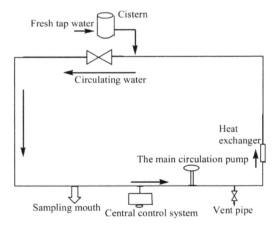

**Figure 1.** Experimental device for the water distribution system.

### 2.2. Experimental Methods

#### 2.2.1. Decomposition of OFL in the Pipe Network

The experimental platform consists of loops with different materials, including the cement lining of ductile iron, polyethylene (PE) materials and stainless-steel materials. The diameter of all of the pipes is 150 mm. Before the experiment, the experimental loop and the drug pump were washed with fresh tap water for about 30 min, and the water was drained after cleaning. After that, the fresh tap water was injected again to start the experiment. The experimental parameters such as the flow rate and water temperature were adjusted to the set value through the central control system. In the experiment that needs to adjust the pH of the water body, the solution pH was adjusted by adding the appropriate sodium hypochlorite or NaOH solution to the pipe network using a drug pump according to the experimental conditions. When all of the experimental conditions reach the set value, a certain amount of sodium hypochlorite solution was added. After the residual chlorine reached the desired value, the formulated and fully dissolved OFL solution was added to make the concentration of OFL in the pipe network be 250 µg/L. After the solution was added to the pipe network, 250 mL brown glass bottles were used for sampling at different reaction times, and the appropriate amount of $Na_2S_2O_3$ was added immediately to terminate the experiment. The concentration of OFL at different times was measured by HPLC at the end of the experiment. We note that while the real OFL concentration can be significantly lower than the concentration (250 µg/L) used in this study, the underlying mechanism/properties of the OFL revealed in this study can be used as guidance to control/manage the OFL in the real water supply system with low concentrations.

### 2.2.2. Degradation of OFL in Beaker Deionized Water

The appropriate amount of deionized water was added to the 250 mL conical bottle, and the appropriate amount of phosphoric acid or NaOH solution was added to adjust the pH. After that, the NaClO reserve liquid was added to reach the residual chlorine requirement. Finally, the OFL reserve liquid was added to start the reaction. At different reaction times, a 1 mL sample was added to a brown reagent flask, prior to which 2.0 mL Agilent with 0.1 mL $Na_2S_2O_3$ was already added into the flask. The concentration of OFL was measured using HPLC.

### *2.3. Analytical Methods*

### 2.3.1. The Analysis Method of Chlorine and OFL

A n,ndiethyl-1,4-phenylenediamine sulfate (DPD) spectrophotometry was used with the DR-2800 spectrophotometer (HACH, Loveland, CO, USA) to measure the concentration of chlorine at a detection wavelength of 530 nm. The concentration of OFL was monitored by an Agilent 1200 high performance liquid chromatography (HPLC) system (Agilent Technologies, Santa Clara, CA, USA) with a UV detector (Shimadzu, Kyoto, Japan) at 293 nm. The separate column was the Zorbax Eclipse XDB-C18 column (4.6 mm × 150 mm, 5 µm particle size) with a column temperature of 30 °C. A mixture of 0.1% formic acid/acetonitrile (75/25, *v/v*) was used as eluents at the flow rate of 1.0 mL/min and the injection volume of 50 µL.

### 2.3.2. LC-MS Analysis Method of OFL Chlorination Products

The chlorination products of OFL were analyzed by the solid phase extraction-liquid chromatography-mass spectrometry (SPE-LC-MS) technique. Before the solid phase extraction, 1 L samples were filtered by an ultrafiltration membrane with an aperture of 0.45 µm to remove large particles. The hydrophilic lipophilic balanced (HLB) extraction cartridges were activated by adding 6 mL and 10 mL of pure water in an orderly way. Next, 1 L samples were loaded into activated cartridges at a flow rate of 5 mL/min. 6 mL methanol was used to elute OFL and its chlorination byproducts, and then the extracts were concentrated to a final volume of 0.1 mL with a gentle nitrogen stream. Finally, the OFL and its intermediates were analyzed by electrospray ionization/liquid chromatograph/mass spectrometry (ESI+/LC/MS) (Agilent 6460, Varian, Palo Alto, CA, USA). A Zorbax SB-18 capillary column with a particle size of 5 µm (2.4 mm × 150 mm, 5 µm) was used. The mobile phase at a flow rate of 0.3 mL/min with 0.2% formic acid (A) and acetonitrile (B) was used. The elution gradient was: 20% B for 5 min, then increased to 35% B over 10 min, followed by a change to 80% B over

5 min, before being finally increased to 100% B over 3 min. An ESI source in positive ion modes with the mass range of the total ion-current (TIC) from 50 to 500 was used in the mass spectrometer analysis. The carrier gas temperature of the ionization conditions of the source was 350 °C. The pressure was 25 psig, theragentor voltage was 110 eV, and the capillarity voltage was 4000 V.

### 2.3.3. The Analytical Method for THMs and HAAs

The liquid extraction was conducted for the pretreatment of the water sample before the analysis of THMs and HAAs. The sample was then placed in a 40 mL glass bottle, and 2 mL MTBE and 8 g anhydrous $Na_2S_2O_4$ were added. The bottle was shocked to dissolve $Na_2S_2O_4$ and was then set for 30 min. After that, 0.2 mL of the upper liquid was extracted into a brown chromatograph bottle for analysis.

Gas chromatography with an electron capture detector (GC-ECD) (GC-450, Varian, Palo Alto, CA, USA) was used to detect the concentration of THMs during the OFL chlorination. An Agilent DB-5 capillary column (30 m × 0.25 mm × 0.25 μm) (Agilent Technologies, Santa Clara, CA, USA) with a flow of 1 mL/min was used in the test. A splitless injection was used at 150 °C in the injectors. The temperature program of the ECD detector was 1 min at 50 °C, before being increased at 20 °C/min to 250 °C.

For the HAA detection, the samples were first derivatized with methyl tert-butyl ether (MTBE). GC (GC-450, Varian, Palo Alto, CA, USA) with an Agilent HP-5 capillary column (15 m × 0.25 mm × 0.25 μm, Agilent Technologies, Santa Clara, CA, USA) was used in the HAAs detection. A headspace injection with a temperature of 175 °C in the injectors was used. The temperature of the detector was 300 °C and the column flow was 1 mL/min. The temperature program was 5 min at 40 °C, increased by 10 °C/min to 140 °C, and then increased by 25 °C/min to 190 °C, and held for 3 min.

## 3. Results and Analysis

### 3.1. Kinetics of the Degradation of OFL by Free Chlorine

#### 3.1.1. Degradation of OFL at Different Free Chlorine Concentrations

The test conditions were: the pH was 7.4, the reaction temperature was 20 °C, the pipe material was PE (PE pipe material has been widely used in China, especially for relatively small diameters, due to its low cost and great ability to prevent corrosion), the flow rate was 1 m/s, the initial concentration of OFL was 250 μg/L, and the initial concentrations of free chlorine were 0.3 mg/L, 0.7 mg/L, 1.0 mg/L and 1.3 mg/L, respectively.

As can be seen from Figure 2a,b, the concentration of OFL gradually decreases with the increase of the reaction time. One can see that the effect of the initial total free chlorine concentration on the degradation of OFL is significant both in the pipe network and deionized water. When the free chlorine concentration is 0.7 mg/L, 1.0 mg/L and 1.3 mg/L, the removal rate of OFL can reach more than 90% after 5 min in the pipe network, while the removal rate can only reach 80% at a free chlorine concentration of 0.3 mg/L. After a 10 min reaction time, when the initial total free chlorine concentration increased from 0.3 mg/L to 1.3 mg/L, the removal rates were 91.7%, 96.8%, 100%, and 100%. In the deionized water, the removal rate of OFL reaches 100% at 3 min when the concentration is 1.3 mg/L. When the initial total free chlorine concentration increased from 0.3 mg/L to 1.3 mg/L, the removal rate was 94.5%, 100%, 100%, and 100%. The comparison shows that the degradation rate of OFL in the deionized water is higher than that in the pipe network. The rationale behind this finding is that organic matter, pipe scale and biofilm in the pipe network consume part of the free chlorine, which reduces the proportion of total free chlorine used for the degradation of OFL in the pipe network.

The experimental data obtained from the degradation of OFL in deionized water at different free chlorine concentrations were fitted with the pseudo first-order reaction Kinetics equation, with results shown in Figure 3. One can see that the linear relationship between $\ln[C/C_0]$ and the reaction time is strong. This indicates that the reaction of the OFL chlorination by free chlorine (OOFC) is a first-order reaction. Figure 4 shows that a strong linear relationship can be observed between the pseudo first-order reaction rate constant ($k_{obs}$) and the free chlorine concentration, indicating that the reaction between the free chlorine concentration and OFL conforms to the second-order kinetic model. $k_{obs}$ is the slope obtained by the linear fitting $\ln[C/C_0]$ and the reaction time T, and a larger $k_{obs}$ value indicates a faster reaction rate.

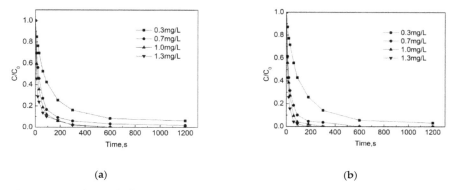

(a)                                                          (b)

**Figure 2.** Degradation of Ofloxacin (OFL) under different free chlorine concentrations in (**a**) the water distribution system and in (**b**) deionized water. [C] represents the concentration of OFL at different reaction times, and [C$_0$] represents the initial OFL concentration.

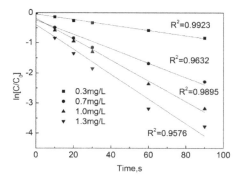

**Figure 3.** Dynamic fitting curves of OFL under different free chlorine concentrations in deionized water.

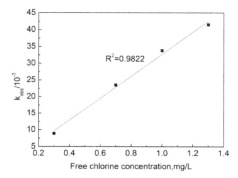

**Figure 4.** The relation curve between the effective free chlorine concentration and $k_{obs}$.

3.1.2. Degradation of OFL at Different pH Conditions

Test conditions: the initial free chlorine concentration was 0.3 mg/L, the initial pipe network concentration of OFL was 250 µg/L, the pipe material was PE, the pipe flow rate was 1 m/s, the temperature was 20 °C, and the degradation of OFL in both the beaker deionized water and in the PE pipe were investigated.

Figure 5a,b shows that the degradation rate of OFL is the fastest under neutral conditions, and that the relative removal efficiency is relatively low under acidic conditions, followed by the lowest rate under alkaline conditions. In the pipe network, after the reaction of 5 min, the removal rates of OFL are 89%, 92.7%, 86.5% and 83.6% at pHs of 6.5, 7.4, 8.0, and 9.0, respectively. In the deionized water, the removal rates of OFL were 80%, 85.7%, 37.7%, and 6.9% at pHs of 6.5, 7.4, 8.0, and 9.0, respectively. One can see that the pH value has a significant effect on the OOFC, and that the degradation rate of OFL in the pipe network is faster than in the deionized water. This may be due to some substances in the water and pipe scale that can promote the degradation of the OFL chlorination. When pH $\leq 7.4$, both the degradation rate of OFL in the pipe network and in the deionized water are faster. When pH = 7.4, the removal efficiency of OFL is the highest, while at pH = 9.0 the removal efficiency of OFL is the lowest. This is because the pH has a combined effect on the presence of sodium hypochlorite and the presence of OFL. The OFL is an amphoteric compound, and the OFL has a different dissociation in water in different pH conditions. The OFL mainly exists in the form of $OFL^+$ under acidic conditions, in the form of $OFL^0$ under neutral conditions, and in the form of $OFL^-$ under alkaline conditions. When the pH is 6.5 and 7.4, the chlorination reaction rate is mainly determined by hypochlorous acid (HOCl) and OFL. When the pH is higher than 7.4 and the concentration of HOCl decreases, the main participation in the reaction is $ClO^-$. As the $ClO^-$ chlorination is weaker than HOCl [17], the reaction rate is reduced.

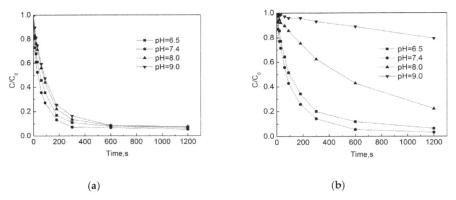

(a)             (b)

**Figure 5.** Degradation of OFL under different pH in (**a**) water distribution system; (**b**) deionized water.

As shown in Figure 6, in the pipe network, when the pH is 7.4, the pseudo first-order rate constant is the highest, with a $k_{obs} = 0.0084$ s$^{-1}$, and when the pH is 9.0 the pseudo first-order rate constant is the lowest, with a $k_{obs} = 0.0061$ s$^{-1}$. In the deionized water, when the pH is 7.4 the pseudo first-order rate constant is the highest, at $k_{obs} = 0.0048$ s$^{-1}$, and when the pH is 9.0 the pseudo first-order rate constant is the lowest, at $k_{obs} = 0.0002$ s$^{-1}$. One can see that there is a significant difference in the degradation rate of OFL chlorination under different pH conditions in the pipe network and deionized water, and the difference in degradation is even greater under alkaline conditions. This is because as the increase of pH, substances in water and pipe scale can promote the degradation of OFL.

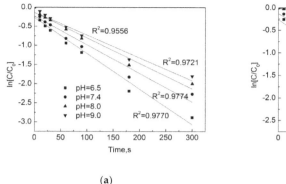 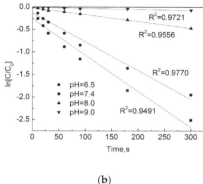

(a)             (b)

**Figure 6.** Dynamic fitting curves of OFL under different pHs in (**a**) the water distribution system, and in (**b**) the deionized water.

3.1.3. Degradation of OFL under Different Pipe Conditions

Test conditions: the initial free chlorine concentration was 0.3 mg/L, the initial OFL concentration in the pipe network was 250 μg/L, the pipe flow rate was 1 m/s, the temperature was 20 °C, the pH was 7.4, and the pipe types included ductile iron pipes, stainless steel pipes and PE pipes. The results are shown in Figure 7.

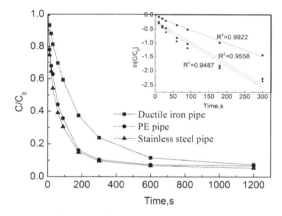

**Figure 7.** Degradation of OFL under different pipe materials in the water distribution system.

Figure 7 shows that after the reaction of 3 min, the removal rates of OFL in the ductile iron pipes, PE and stainless-steel pipes were 62.9%, 83.9%, and 85.1%, respectively. After the reaction of 20 min, the removal rates of OFL in the ductile iron pipes, PE and stainless-steel pipes were 92.7%, 93.2% and 94.6%, respectively. The pseudo first-order rate constants of OFL in the ductile iron pipes, PE and stainless-steel pipes are $k_{obs}$ = 0.0048 s$^{-1}$, 0.0075 s$^{-1}$ and 0.0076 s$^{-1}$, respectively. One can see that the degradation rate of OFL in the ductile iron pipes is the slowest, while the rate of degradation in the PE pipes is similar to that in the stainless-steel pipes. By observing the visual pipe segment within the pipe network (i.e., the transparent segment of each pipe), the pipe scale in the stainless-steel pipe network is the highest, followed by the PE pipe, and the ductile iron pipe is the lowest. The analysis of the water in the three pipe networks shows that there is a small amount of iron ions in the water of the stainless-steel pipe network. According to the previous study [18], Fe$^{2+}$ could react with free chlorine to produce Fe$^{3+}$, and the chlorination process produces some free radicals [19]. This might be one reason that results in a faster degradation of OFL in the stainless-steel pipe compared to the one in the other two pipes. By considering the effect of iron ions and the pipe scale, the degradation rate of OFL in the PE pipes is approximately equal to that in the stainless-steel pipes, while that in the ductile cast iron pipe network is the slowest.

3.1.4. Degradation of OFL at Different Flow Rates

Since the effect of different flow rates on the degradation of OFL is only present in the pipe network, the PE pipes were chosen and the degradation in the deionized water is not considered. Therefore, in order to study the effect of different flow rates on the degradation of OFL, the initial free chlorine concentration was 0.3 mg/L, the concentration of OFL in the initial pipe network was 250 µg/L, the temperature was 20 °C, the pH of the pipe network was 7.4, and the flow rates were 0.5 m/s, 1.0 m/s, and 1.5 m/s.

Figure 8 indicates that the degradation rate of OFL in the pipe network with the flow rate of 1.0 m/s is the fastest, while that in the pipe network with the flow rate of 1.5 m/s is the slowest. The removal rate of OFL was 90.8%, 89.0% and 87.8% at the flow rates of 1.0 m/s, 1.5 m/s and 1.0 m/s, respectively. The effect of the increase in the flow rate on the reaction rate is not obvious. According to hydraulics knowledge, when the speed is 0.5 m/s, the Reynolds number is higher than 2300 and the water is already in turbulent state. When the flow rate increases from 0.5 m/s to 1.5 m/s, the material exchange is sufficient, and hence the effect of the increase of the flow rate on the degradation rate of OFL is insignificant.

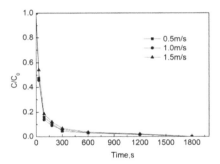

**Figure 8.** Degradation of OFL under different flow rates in the water distribution system.

3.1.5. Degradation of OFL at Different Temperatures

Test conditions: the initial free chlorine concentration was 0.3 mg/L, the initial pipe network concentration of OFL was 250 µg/L, the pipe material was PE, the pipe flow rate was 1 m/s, the pH of the water in the pipe network was 7.4, and the temperatures were 15 °C, 20 °C and 25 °C. The results are shown in Figure 9.

Figure 9 shows that the temperature has a significant effect on the degradation of OFL both in the pipe network and in the deionized water. The degradation rate continues to rise as the temperature increases from 15 °C to 20 °C. After 90 s, the removal rates of OFL in the pipe network were 48.6%, 57.2%, and 63.9%, and those in the deionized water were 54.2%, 64.4%, and 74.6%. As shown in Figure 10, as the temperature increases, the degradation rate in both conditions increases. This is because, when the temperature rises, the average kinetic energy and activation molecular weight of the molecule in the reaction system will increase, resulting in an increase in the number of effective collisions of molecules during the reaction process, as a result of which the chlorination reaction rate will be accelerated. However, we also note that at the same temperature, the degradation rate of OFL in the pipe network is slightly lower than that in the deionized water. In the pipe network, when the reaction temperature rises from 15 °C to 25 °C, the degradation rate of OFL increases from 0.0057 s$^{-1}$

to 0.0085 s$^{-1}$, while it rises from 0.0059 s$^{-1}$ to 0.0099 s$^{-1}$ in the deionized water. This is due to the water organic matter, pipe scale, and pipe wall microorganisms, which consume a certain amount of free chlorine and reduce the free chlorine for the OFL chlorination.

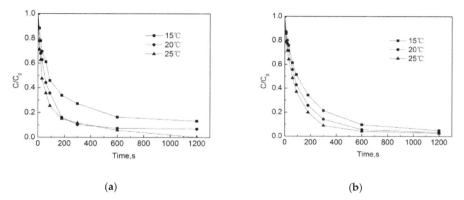

(a)                                                            (b)

**Figure 9.** Degradation of OFL under different temperatures in (**a**) the water distribution system, and in (**b**) the deionized water.

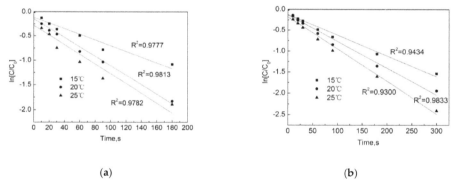

(a)                                                            (b)

**Figure 10.** Dynamic fitting curves of OFL under different temperatures in (**a**) the water distribution system, and in (**b**) the deionized water.

To find out the effect of the temperature on the reaction, ln$k$ and 1/$T$ were fitted according to the Arrhenius equation, and the results are shown in Figure 11. According to the Arrhenius equation:

$$\ln k = -\frac{Ea}{RT} + \ln A \tag{1}$$

where $A$ is the frequency factor, $Ea$ is the reaction activation energy, $R$ is the gas general constant with a value of 8.314, and $T$ is the thermodynamic temperature. According to the equation and the slope of the fitting line, the reaction activation energy of the OOFC in the pipe network is 37.04 kJ/mol, while it is 28.63 kJ/mol in the deionized water. This means that the degradation of OFL in the deionized water is more likely to occur.

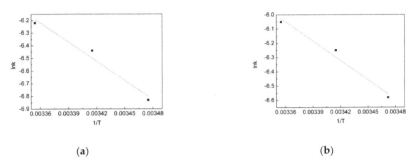

(a)                                                              (b)

**Figure 11.** Arrhenius fitting of OFL chlorination by free chlorine (OOFC) in (**a**) the water distribution system, and in (**b**) the deionized water.

*3.2. Analysis of the Production Law of Intermediary Products.*

During the intermediate product analysis, the experimental conditions are as follows: the pipe material was PE, the flow rate was 1.0 m/s, the temperature was 20 °C, the pH was 7.4, the initial free chlorine concentration was 0.3 mg/L, and the initial OFL concentration was 250 μg/L. At 0 h, 15 min, 1 h, 3 h, 6 h and 24 h after the experiment started, a 1 L sample was added to the glass bottle respectively and 2 g of $Na_2S_2O_4$ were added immediately to terminate the experiment. The results of the chromatogram at reaction times of 0 h and 24 h are shown in Figure 12, and the mass spectrogram of the products is given in Figure 13. Based on the information presented in Figure 13, the mass-charge ratio and molecular weight of the main products are derived as shown in Table 1. The potential degradation path is derived as given in Figure 14, with a detailed analysis given below.

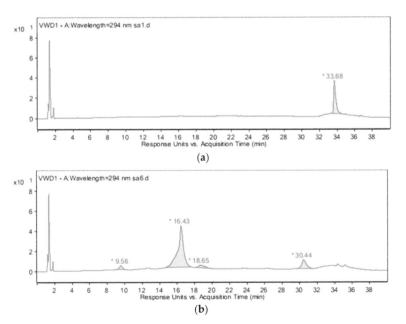

(a)

(b)

**Figure 12.** The chromatogram of the water sample at reaction times of: (**a**) 0 h and (**b**) 24 h.

By comparing with the 0 h water sample, one can see that the reaction has a relatively obvious fragmention peak at 16.45 min, 18.67 min, and 30.42 min. By comparison, one can see that the fragment ion can be found at 15 min, 1 h, 3 h, 6 h and 24 h and that the molecular weight is 261.1. At the beginning of the reaction (Figure 14), the fragment ions attack the hydroxyl group on the piperazine ring, seize the hydrogen and lose one $H_2O$ molecule to form the chlorination product M-18 (the charge-to-charge ratio is 345.2), also losing a $CO_2$ molecule to form the chlorination product M-44 (mass ratio is 318).

After further analysis of its chlorination products, it is found that the piperazine ring is one of the main groups involved in the reaction, and the main point involved in the chlorination reaction is the N4 atom on the piperazine ring. It loses one $-CH_3$ to form the chlorination product M-61(the mass ratio is 301), continues to act on the N4 atom on the piperazine ring, and removes $C_3H_7N$ to form the chlorination product M-101 (the mass ratio is 261.1). The chlorination product M-139 is formed by the ring opening of oxazine (the charge ratio is 223.1). The N2 atom on the quinolone ring is replaced by Cl to form the chlorination product M-108 (the charge-to-charge ratio is 254.1).

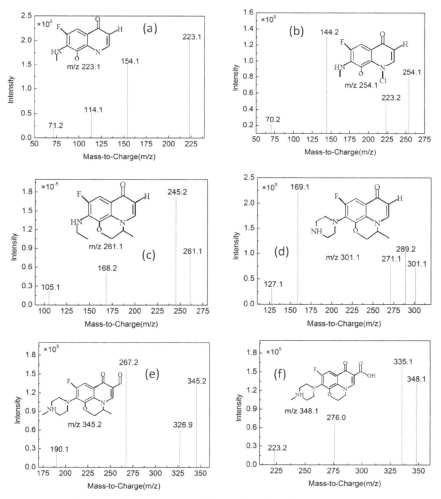

**Figure 13.** Mass spectrogram of the products (for details see Table 1).

**Table 1.** Mass–charge ratio and molecular weight of the main products.

| Name Abbreviations | Mass–Charge Ratio | Molecular Weight | Production Structure |
|---|---|---|---|
| 1. M-139 | 223.1 | 223.1 | 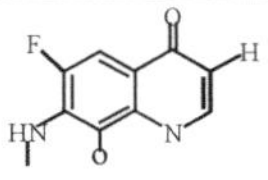 |
| 2. M-108 | 254.1 | 254.1 | 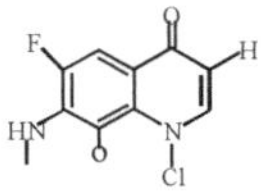 |
| 3. M-101 | 261.1 | 261.1 | 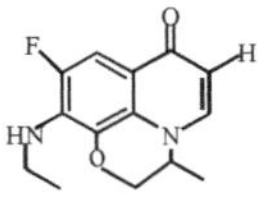 |
| 4. M-61 | 301.1 | 301.1 | 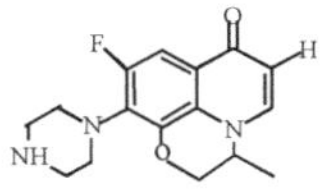 |
| 5. M-17 | 345.2 | 345.2 | 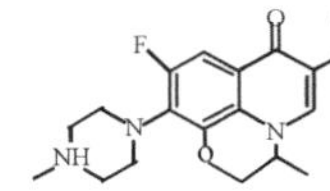 |
| 6. M-12 | 348.1 | 348.1 | 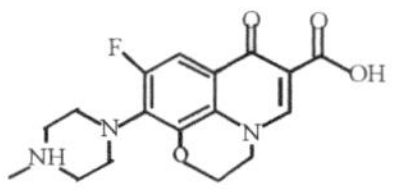 |
| 7. M+14 | 386.1 | 386.1 | 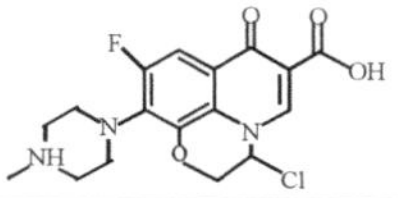 |

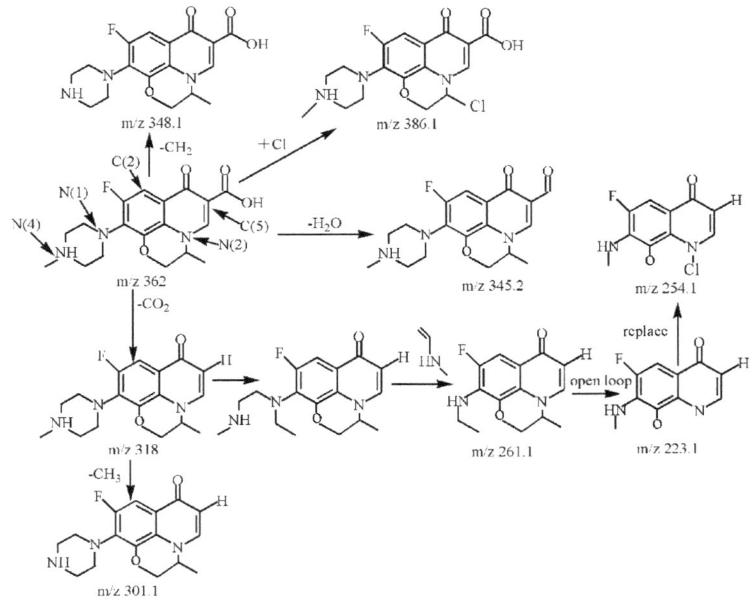

**Figure 14.** Degradation path of OFL.

*3.3. Analysis of the Formation Law of THMs and HAAs*

The experimental conditions are as follows: the pipe material was PE, the flow rate was 1.0 m/s, the temperature was 20 °C, the pH was 7.4, the initial free chlorine concentration was 0.3 mg/L, and the initial OFL concentration was 250 µg/L. At 0 h, 15 min, 1 h, 3 h, 6 h, and 24 h, a 200 mL sample was taken into the glass bottle respectively, and 1 g of $Na_2S_2O_4$ was immediately added to terminate the experiment. Then, a liquid extraction was used to preprocess the water sample, before a 30 mL sample were placed in a 40 mL glass bottle. After that, 2 mL of MTBE and 8 g of anhydrous $Na_2S_2O_4$ was added, shaking to dissolve $Na_2S_2O_4$, and the bottle was left for 30 min. Finally, 0.2 mL of the upper liquid was extracted to a brown chromatograph bottle for the analysis.

The formation curves of three kinds of THMs detected in the experiment are shown in Figure 15, which are trichloromethane (TCM), monobromodichloride (DCBM) and dibromodichloride (DBCM). It can be seen that THMs in the raw water mainly exist in the form of TCM. The concentration of three THMs increases significantly with the increase of the reaction time in the first 6 h. After 6 h, the concentration of DCBM and DBCM increases slowly while the concentration of TCM still increases significantly. After 24 h, the concentrations of TCM, DCBM and DBCM are 36.61 µg/L, 27.23 µg/L, and 5.808 µg/L, respectively.

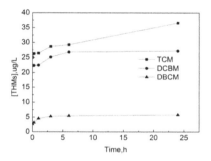

**Figure 15.** The formation curve of trihalomethanes (THMs) in the process of OOFC.

As can be seen from Figure 16, HAAs in water bodies mainly exist in the form of monochloroacetic acid (MCAA). After the reaction begins, the concentration of free chlorine in the pipe network is relatively high, the free chlorine components used for the reaction with OFL are sufficient and the concentration of three HAAs gradually increases with the increase of the reaction time. The increase of MCAA is particularly obvious, and the increase of monobromoacetic acid (MBAA) and dichloroacetic acid (DCAA) are subsequently stable. The rate of the production of MCAA is the fastest in the first 6 h, and due to the continuous consumption of free chlorine the rate of production decreases later. After 24 h, the concentrations of MCAA, MBAA and DCAA are 1.32 µg/L, 0.23 µg/L, and 0.08 µg/L, respectively.

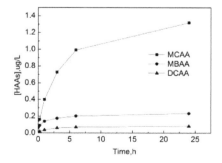

**Figure 16.** The formation curve of haloacetic acids (HAAs) in the process of OOFC.

## 4. Conclusion

1.  The reaction rate of OFL in the pipe networks and deionized water increases with the increase of the free chlorine concentration and temperature. The degradation rate of OFL under the neutral conditions is the fastest. The removal efficiency is lower under acidic conditions, and the removal efficiency is the lowest under alkaline conditions.

2.  The degradation of OFL in the pipe network does not change significantly with the different flow rates, while it is affected by different pipe materials. Under the combined effect of the iron ions and pipe scale, the degradation rate of OFL in the stainless-steel pipe is similar to that in the PE pipe, but both are greater than the ductile cast iron pipe.

3.  As the reaction time increases, the concentrations of THMs and HAAs will gradually increase. THMs mainly exist in the form of TCM, and HAAs mainly exist in the form of MCAA.

4.  Further analysis of the intermediate chlorination product shows that the piperazine ring is one of the main groups involved in the reaction, and that the main point involved in the chlorination reaction is the N4 atom on the piperazine ring, which is mainly responsible for dealkylation and hydroxylation, and which produces intermediate chlorination products.

**Author Contributions:** Writing—original draft, review and editing, W.B.; data collection, formal analysis, investigation and methodology, Y.J.; supervision, H.W.

**Funding:** This research was funded by National Natural Science Foundation of China, grand number 51808497 and Zhejiang Provincial Department of Education General Research Foundation (Natural Science), grand number Y201636517.

**Conflicts of Interest:** The authors declare no conflict of interest.

## References

1.  Janecko, N.; Pokludova, L.; Blahova, J.; Svobodova, Z.; Literak, I. Implications of fluoroquinolone contamination for the aquqtic environment—A review. *Environ. Toxicol. Chem.* **2016**, *35*, 2647–2656. [CrossRef] [PubMed]

2.  Snowberger, S.; Adejumo, H.; He, K.; Mangalgiri, K.P.; Hopanna, M.; Soares, A.D.; Blaney, L. Direct photolysis of fluoroquinolone antibiotics at 253.7 nm: Specific reaction kinetics and formation of equally potent fluoroquinolone antibiotics. *Environ. Sci. Technol.* **2016**, *50*, 9533–9542. [CrossRef] [PubMed]

3.  Serna-Galvis, E.A.; Berrio-Perlaza, K.E.; Torres-Palma, R.A. Electrochemical treatment of penicillin, cephalosporin, and fluoroquinolone antibiotics via active chlorine: Evaluation of antimicrobial activity, toxicity, matrix, and their correlation with the degradation pathways. *Environ. Sci. Pollut. Res.* **2017**, *24*, 23771–23782. [CrossRef] [PubMed]

4.  El Najjar, N.H.; Deborde, M.; Journel, R.; Leitner, N.K. Aqueous chlorination of levofloxacin: Kinetic and mechanistic study, transformation product identification and toxicity. *Water Res.* **2013**, *47*, 121–129. [CrossRef] [PubMed]

5.  Liu, X.; Wang, Z.; Wang, X.L.; Li, Z.; Yang, C.; Li, E.H.; Wei, M.H. Status of antibiotic contamination and ecological risks assessment in Chinese Several typical surface-water environment. *Environ. Sci.* **2019**, *5*, 1–13. (In Chinese)

6.  Wang, X.; Li, Y.; Li, R.; Yang, H.; Zhou, B.; Wang, X.; Xie, Y. Comparison of chlorination behaviors between norfloxacin and ofloxacin: Reaction kinetics, oxidation products and reaction pathways. *Chemosphere* **2019**, *215*, 124–132. [CrossRef] [PubMed]

7.  Yassine, M.H.; Rifai, A.; Hoteit, M.; Mazellier, P. Study of the degradation process of ofloxacin with free chlorine by using ESI-LCMSMS: Kinetic study, by-products formation pathways and fragmentation mechanisms. *Chemosphere* **2017**, *189*, 46–54. [CrossRef] [PubMed]

8.  Zhu, L.; Santiago-Schübel, B.; Xiao, H.; Hollert, H.; Kueppers, S. Electrochemical oxidation of fluoroquinolone antibiotics: Mechanism, residual antibacterial activity and toxicity change. *Water Res.* **2016**, *102*, 52–62. [CrossRef] [PubMed]

9.  Ding, C.S.; Zou, B.W.; Miao, J.; Fu, Y.P.; Shen, J.C. Formation process of nitrogenous disinfection byproduct trichloronitromethane in drinking water and its influencing factors. *Environ. Sci.* **2013**, *34*, 3113–3118. (In Chinese)
10. Tay, K.S.; Madehi, N. Ozonation of ofloxacin in water: By-products, degradation pathway and ecotoxicity assessment. *Sci. Total Environ.* **2015**, *520*, 23–31. [CrossRef] [PubMed]
11. Lin, Y.Z.; Liu, X.Y. Research progress of chlorination and disinfection by-products in drinking water. *China Resour. Compr. Util.* **2017**, *35*, 128–130. (In Chinese)
12. Frade, V.M.; Dias, M.; Teixeira, A.C.; Palma, M.S. Environmental contamination by fluoroquinolones. *Braz. Pharm. Sci.* **2014**, *50*, 41–54. [CrossRef]
13. Zhang, Q.; Zheng, F.; Duan, H.-F.; Jia, Y.; Zhang, T.; Guo, X. Efficient numerical approach for simultaneous calibration of pipe roughness coefficients and nodal demands for water distribution systems. *J. Water Resour. Plan. Manag.* **2018**, *144*, 04018063. [CrossRef]
14. Qi, Z.; Zheng, F.; Guo, D.; Maier, H.R.; Zhang, T.; Yu, T.; Shao, Y. Better understanding of the capacity of pressure sensor systems to detect pipe burst within water distribution networks. *J. Water Resour. Plan. Manag.* **2018**, *144*, 04018035. [CrossRef]
15. Qi, Z.; Zheng, F.; Guo, D.; Zhang, T.; Shao, Y.; Yu, T.; Zhang, K.; Maier, H.R. A Comprehensive framework to evaluate hydraulic and water quality impacts of pipe breaks on water distribution systems. *Water Resour. Res.* **2018**, *54*, 8174–8195. [CrossRef]
16. Zheng, F.; Zecchin, A.C.; Newman, J.P.; Maier, H.R.; Dandy, G.C. An adaptive convergence-trajectory controlled ant colony optimization algorithm with application to water distribution system design problems. *IEEE Trans. Evol. Comput.* **2017**, *21*, 773–791. [CrossRef]
17. Zhang, S.; Wang, X.; Yang, H.; Xie, Y.F. Chlorination of oxybenzone: Kinetics, transformation, disinfection byproducts formation, and genotoxicity changes. *Chemosphere* **2016**, *154*, 521–527. [CrossRef] [PubMed]
18. Pérez-Moya, M.; Graells, M.; Castells, G.; Amigó, J.; Ortega, E.; Buhigas, G.; Pérez, L.M.; Mansilla, H.D. Characterization of the degradation performance of the sulfamethazine antibiotic by photo-Fenton process. *Water Res.* **2010**, *44*, 2533–2540. [CrossRef] [PubMed]
19. Li, C.; Wang, Z.; Yang, Y.J.; Liu, J.; Mao, X.; Zhang, Y. Transformation of bisphenol in water distribution systems: A pilot-scale study. *Chemosphere* **2015**, *125*, 86–93. [CrossRef] [PubMed]

*Article*

# Regional and Seasonal Distributions of *N*-Nitrosodimethylamine (NDMA) Concentrations in Chlorinated Drinking Water Distribution Systems in Korea

Sunyoung Park [1], Sungjin Jung [1] and Hekap Kim [2,*]

[1] Department of Environmental Science, Kangwon National University, Chuncheon, Gangwon-do 24341, Korea; tj54795@naver.com (S.P.); jsjbin878@naver.com (S.J.)
[2] School of Natural Resources and Environmental Science, Kangwon National University, Chuncheon, Gangwon-do 24341, Korea
* Correspondence: kimh@kangwon.ac.kr; Tel.: +82-33-250-8577

Received: 25 November 2019; Accepted: 13 December 2019; Published: 14 December 2019

**Abstract:** Volatile *N*-Nitrosamines (NAs), including *N*-nitrosodimethylamine (NDMA), an emerging contaminant in drinking water, have been reported to induce cancer in animal studies. This study aims to investigate the regional and seasonal distributions of the concentrations of NDMA, one of the most commonly found NAs with high carcinogenicity, in municipal tap water in Korea. NDMA in water samples was quantitatively determined using high-performance liquid chromatography-fluorescence detection (HPLC-FLD) as a 5-dimethylamino-1-naphthalenesulfonyl (dansyl) derivative after optimization to dry the SPE adsorbent and remove dimethylamine prior to derivatization. Tap water samples were collected from 41 sites in Korea, each of which was visited once in summer and once in winter. The average (±standard deviation) NDMA concentration among all the sites was 46.6 (±22.7) ng/L, ranging from <0.13 to 80.7 ng/L. Significant NDMA differences in the regions, excluding the Jeju region, were not found, whereas the average NDMA concentration was statistically higher in winter than in summer. A multiple regression analysis for the entire data set indicated a negative relationship between NDMA concentration and water temperature. High levels of NDMA in Korea may pose excessive cancer risks from the consumption of such drinking water.

**Keywords:** cancer risk; distribution; drinking water; *N*-nitrosodimethylamine; regional; seasonal

## 1. Introduction

*N*-Nitrosamines (NAs) in municipal drinking water have been reported as a group of disinfection byproducts (DBPs) formed by chloramination (chlorination of water containing ammonia or nitrite) and ozonation [1]. Among them, six NAs, including *N*-nitrosodimethylamine (NDMA), *N*-nitrosomethylethylamine (NMEA), *N*-nitrosodiethylamine (NDEA), *N*-nitrosodipropylamine (NDPA), *N*-nitrosodibutylamine (NDBA), and *N*-nitrosopyrrolidine (NPYR) have been classified as probable human carcinogens (group B2) and designated as contaminants under the second Unregulated Contaminant Monitoring Rule (UCMR 2) by the United States Environmental Protection Agency (EPA) [2].

In particular, NDMA has been of the greatest concern because it is the most frequently found NA in drinking water and poses an extremely high carcinogenic risk from oral exposure with a drinking water unit risk (DWUR) of $1.4 \times 10^{-3}$ per μg/L [3]. This value is ~1000 times greater than those of bromodichloromethane ($1.8 \times 10^{-6}$ per μg/L) [4] and dichloroacetic acid ($1.4 \times 10^{-6}$ per μg/L) [5], which are frequently found at μg/L levels in chlorinated drinking water. However, NDEA has been found much less frequently and in lower concentrations than NDMA [6,7], even though its carcinogenic potential is reported to be the greatest among the NAs (DWUR of $4.3 \times 10^{-3}$ per µg/L) [8]. Furthermore, diethylamine (DEA), which is a major organic precursor to NDEA, is rarely found in drinking water or surface water samples in the study sites [9]. Therefore, this study is limited to NDMA.

NDMA concentrations vary significantly according to region and season. A nationwide survey of source and finished water samples collected in Japan indicated that NDMA concentrations were lower in finished water samples than those in source water samples and that those in the finished water samples were higher in winter than in summer [10]. However, NDMA was detected to be above the method detection limit (MDL) in only a few finished samples (25.4%, 15/59) and had a maximum concentration of 10 ng/L [10]. To the contrary, a study conducted in Spain [11] indicated that NDMA concentration increased throughout the distribution system following chlorination. This study also indicated higher NDMA concentrations in winter than in summer and fall (3.2–20 ng/L vs. 1.5–2.5 ng/L and 0.89–9.2 ng/L in the distribution system, respectively) [11]. A nationwide survey conducted in Korea from 2013 to 2015 indicated a very low detection rate (8.1%) and a maximum concentration of 13.0 ng/L [12]. Studies conducted in China found similar values in finished water and tap water samples with a maximum concentration of 13.9 ng/L [13,14]. However, some surveys carried out in Canada and the United States revealed much higher concentrations of NDMA in finished water and tap water (up to 630 ng/L) than the other aforementioned studies [15–17].

The NDMA concentrations measured in the plants may differ significantly from those of the water in the faucets from which people consume water supplied through water distribution systems. Higher concentrations of NDMA typically corresponded with increased residence time in distribution systems, and higher NDMA concentrations were observed in the faucets than in the plants [11,18,19]. Therefore, from a public health perspective, it would be beneficial to measure the concentrations of NDMA in drinking water collected from the faucets from which water is consumed by the public with/without additional processing (e.g., heating/boiling and further treatment through water purifiers). Nonetheless, human health risk assessments for most DBPs including NDMA are commonly conducted using the data obtained for water at treatment plants. This may lead to the underestimation of possible health effects caused by the exposure via oral consumption, inhalation, and dermal absorption. This is also true of Korea, where it appears that NDMA is not a great concern because of low excess cancer risks ($<10^{-6}$) estimated using the treatment plant data.

NDMA concentrations vary according to the season and are typically higher in winter than in other seasons [10,11,18]. A higher concentration of NDMA in winter appears to be attributed to less effective pre-chlorination, which is used to destroy NDMA precursors, including dimethylamine (DMA), at lower temperatures [19].

To quantitatively determine the concentration of NDMA in water, pretreatment procedures, including NDMA adsorption onto solid particles, sorbent drying, and solvent elution steps, are crucial. Aqueous NDMA is commonly adsorbed onto coconut charcoal [20,21], Ambersorb 572 [6,17], and Carboxen 572 [22]. The compound is then eluted with an organic solvent, such as dichloromethane (DCM) or a mixture of DCM and methanol, following adsorbent drying under vacuum or purging with inert gas (e.g., $N_2$). The eluate is concentrated and then analyzed for NDMA using either gas chromatography (GC) or high-performance liquid chromatography (HPLC) following derivatization.

In the above procedures, the chosen adsorbent and drying method are critical. NDMA might form on the surface of activated carbon from secondary amines in the presence of nitrogen and oxygen during sample loading and adsorbent drying [23,24]. This suggests that the rapid drying of the adsorbent with or without minimal contact with atmospheric gases is necessary to avoid the artifact formation of NDMA during the pretreatment process. Adsorbent drying is typically performed with a vacuum pump [20,22,25,26] or nitrogen gas flow [14,21]. However, detailed descriptions of vacuum pressure, flow rate, and/or drying time have not been made. Thus, excessive or insufficient drying may lead to the loss or incomplete extraction of NDMA, respectively.

When HPLC coupled with fluorescence detection (FLD) is used to determine NAs, including NDMA, they are most frequently converted to fluorescent 5-dimethylamino-1-naphthalenesulfonyl (dansyl) derivatives [22,27–29]. However, this method requires the removal of residual secondary amines, including DMA, which are organic precursors to NAs from the eluate prior to derivatization, because drinking water typically contains a few μg/L levels of secondary amines [9,30,31]. Otherwise, NA concentrations can be overestimated because of the additional formation of dansyl derivatives as a result of the reactions of secondary amines contained in the water itself, which are not formed from NAs by denitrosation with dansyl chloride.

In this study, an analytical method for determining NDMA in water using HPLC-FLD is established following the optimization of sorbent-drying conditions and the removal of secondary amines. Thereafter, the regional and seasonal distributions of the NDMA concentrations in tap water samples collected nationwide from 41 sites in Korea in summer and winter are investigated.

## 2. Materials and Methods

### 2.1. Chemicals and Reagents

NDMA, *N*-nitrosomethylbutylamine (NMBA, a surrogate), DMA, sodium thiosulfate ($Na_2S_2O_3$), and Carboxen® 572 were purchased from Sigma-Aldrich (St. Louis, MO, USA). Acetone, acetonitrile ($CH_3CN$), DCM, and methanol were obtained from Honeywell Burdick & Jackson (Muskegon, MI, USA), and NaOH, $NaHCO_3$, $Na_2SO_4$, and glacial acetic acid were purchased from Daejeong Chemicals & Metals (Siheung, Gyeonggi-do, Korea). Dansyl chloride and 48% hydrobromic acid were purchased from Calbiochem (San Diego, CA, USA) and Wako Pure Chemical Industries, Ltd. (Osaka, Gansai, Japan), respectively. Silica gel blue was purchased from Showa (Saitama, Japan).

A reagent for the denitrosation of NDMA to dimethylamine was prepared by diluting 1 mL of a 48% HBr solution to 10 mL with glacial acetic acid. A dansylating reagent was made by dissolving 25 mg of dansyl chloride in acetone and diluting it to 50 mL. A pH 10.5 buffer solution was prepared by dissolving 0.6 g of NaOH and 2.0 g $NaHCO_3$ in water and diluting it to 50 mL. All reagents were stored in a refrigerator at 4 °C and used within 2 weeks.

### 2.2. Optimization of the Analytical Method for NDMA in Water

The analytical method used in this study was an improved modification of those in previous reports in which dry air was not used to remove water from the cartridge [6,25,26], and secondary amines, including DMA, were not removed prior to chemical derivatization [22]. Optimizations were carried out to dry the sorbent cartridge, select the extracting solvents, and remove secondary amines. A 500 mL water sample spiked with 2 μL of a surrogate (NMBA, 5 mg/L) was passed through a cartridge containing 2.0 g of Carboxen® 572 adsorbent at a flow rate of 10 mL/min. The cartridge was mounted on a vacuum manifold (Visiprep™ SPE vacuum manifold, Sigma-Aldrich) and dried under vacuum (−30 kPa) with and without a silica gel trap (16 g in an impinger) for 60 min. Another drying method used a 1-L/min nitrogen gas flow through the cartridge for 60 min. NDMA in the adsorbent was eluted with 15 mL of a solvent system at a flow rate of 10 mL/min. Two eluent options were examined: (1) a mixture of DCM and methanol (95:5, v/v) and (2) DCM only. Each experimental setting was repeated three times. After the above optimizations were conducted, the duration required for adsorbent drying was tested at 10, 30, 60, and 90 min.

DMA removal efficiency was tested by spiking 0.5 μL of its stock solution (1000 mg/L) or its 1 mL concentrate into the 15 mL eluate. In the former method, the 15 mL eluate was transferred to a 40 mL vial, to which 3 mL of a 1 N HCl solution was added. The mixture was vigorously shaken for 10 min using a mechanical shaker (SR-2DS, Taitec; Koshigaya, Japan). The organic layer was separated and concentrated to 1 mL using a gentle nitrogen gas stream after drying over approximately 0.5 g of $Na_2SO_4$. In the second method, the 15 mL eluate in a centrifuge tube was concentrated to 1 mL using a mild nitrogen gas stream, and 1 mL of 1 N HCl was then added to the concentrate. The mixture was shaken for 5 min using a Maxi Mix II vortex mixer (Barnstead Thermolyne, Dubuque,

IA, USA). The organic phase was separated and dried over approximately 50 mg of $Na_2SO_4$. Both sets of experiments were performed in triplicate.

In a centrifuge tube, 150 μL of the denitrosating reagent was added to each 1 mL concentrate, and the resulting mixture was vortex-mixed for 10 seconds. The mixture was heated at 40 °C for 30 min and concentrated to dryness using a nitrogen gas stream. The pH 10.5 buffer solution (150 μL) and the dansylating reagent (150 μL) were added to the concentrate, and the mixture was shaken for 10 s using the vortex mixer. After the centrifuge tube was heated at 40 °C for 30 min, the mixture was mixed with 50 μL of reagent water. Forty microliter of the analytical sample was injected into the HPLC-FLD system. The instrumental conditions are presented in Table 1.

**Table 1.** High-performance liquid chromatography-fluorescence detection (HPLC-FLD) conditions.

| Parameter | Model/Condition |
|---|---|
| Pump | 515 HPLC pump (Waters Co., Milford, MA, USA) |
| Sample Injection | 717 Plus Autosampler (Waters) |
| Stationary Phase | Skypak C18 (4.6 mm × 250 mm × 5 μm, SK Chemicals, Seongnam, Gyeonggi-do, Korea) |
| Mobile Phase | Water:$CH_3CN$ (45:55, v/v) |
| Detector | 474 Fluorescence detector (Waters) |
| Excitation and Emission Wavelengths | 340 nm and 530 nm |
| Injection Volume | 40 μL |
| Flow Rate | 1 mL/min |

*2.3. Method Validation*

The optimized method was validated for the method detection limit (MDL), the limit of quantitation (LOQ), the linearity of a calibration curve ($r^2$), accuracy, and precision. The MDL and LOQ were estimated according to the US EPA's procedure [25]. A five-point linear calibration curve was drawn using a set of standards with 2.0, 20, 40, 60, and 80 ng/L concentrations, and the coefficient of determination ($r^2$) was calculated to determine the linearity of the calibration curve. The accuracy of both the percent recoveries and percent errors between experimental and theoretical values at three levels (2.0, 30, and 60 ng/L) was evaluated. Precision was evaluated using repeatability expressed as relative standard deviations (RSDs) of the three replicates at the same levels as the accuracy.

*2.4. Sampling and Analysis of Tap Water Samples*

Tap water samples were collected in 250 mL amber glass bottles containing approximately 25 mg of sodium thiosulfate from water faucets at 41 sites distributed nationwide throughout Korea, including Jeju Island (Figure 1). The sites were classified into seven regional groups: Seoul (five sites); Gyeonggi (five sites), including Incheon; Gangwon (five sites); Chungcheong (eight sites), including Daejeon and Sejong; Jeolla (seven sites), including Gwangju; Gyeongsang (nine sites), including Busan, Ulsan, and Daegu; and Jeju (two sites). Eleven of the 41 sites (26.8%) were supplied with drinking water from water treatment plants equipped with advanced oxidation processes (AOPs), such as ozone and powdered-activated carbon (PAC) treatment, including all Seoul sites, one site in Gangwon, and five sites in Gyeongsang. Samples were collected from the same sites in summer (16 August–29 September 2016) and winter (6 January–9 February 2017). On each visit, free and total residual chlorine concentrations, pH, and water temperature were measured in situ.

Secondary amines were determined using GC-MS according to the method by Park et al. [9]. Dissolved organic carbon (DOC) concentrations were measured using a TOC analyzer (Sievers 5310C, Boulder, CO, USA) after samples were filtered through a 0.45 μm membrane filter. Absorbance values at 254 nm ($UV_{254}$) were determined using a UV-Vis spectrophotometer (UV-9100, Human Co., Seoul, Korea). Specific UV absorbance values at 254 nm ($SUVA_{254}$) were calculated by dividing $UV_{254}$ values by DOC concentrations. In addition, total nitrogen concentrations were measured using the persulfate oxidation method [32]. Nitrate concentrations were determined using ion chromatography.

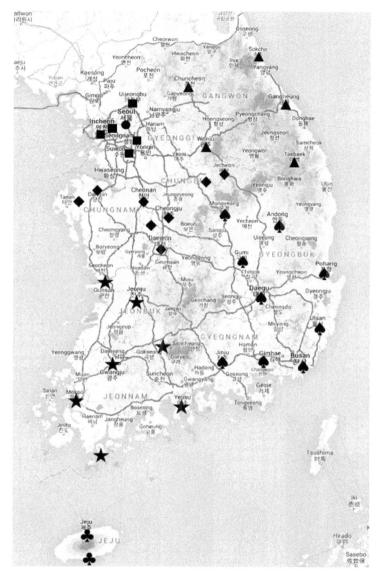

**Figure 1.** Tap water sampling sites. A total of 41 sites [five in Seoul (●), five in Gyeonggi (■), five in Gangwon (▲), eight in Chungcheong (◆), seven in Jeolla (★), nine in Gyeongsang (♠), and two in Jeju (♣)] were visited once in summer and once in winter.

*2.5. Data Analysis*

Data analyses were conducted using IBM SPSS 24 software (Armonk, NY, USA), and a significance level of 5% was used to determine the statistical significance of the tests. A paired *t*-test was conducted to examine differences in NDMA concentrations between summer and winter. Regional variations of NDMA concentrations among the seven regions were tested using a one-way analysis of variance (ANOVA). Causal relationships between water quality parameters (DMA concentrations;

free, combined, and total residual chlorine concentrations; pH; water temperature; DOC concentration; $SUVA_{254}$; total nitrogen concentration; and nitrate concentration) and NDMA concentrations were examined by multiple regression analyses.

## 3. Results and Discussion

### 3.1. Method Optimization

Figure 2 illustrates peak area ratios for six different combinations of three sorbent drying methods and two eluents. Vacuum drying using an in-line silica gel trap and subsequent elution with DCM (first set) indicated the highest chromatographic response; therefore, this set was selected as the first priority for the method. The value for the use of the mixture (5% methanol in DCM, the second set) was approximately half that of DCM only. Vacuum drying without a silica gel trap (third and fourth sets) resulted in lower values than those of vacuum drying with a silica gel trap (first and second sets), probably because of the incomplete drying of the adsorbent because of the withdrawal of atmospheric moisture into the cartridge. This can be explained by the dissolution of NDMA on the adsorbent with an extracting solvent. Adsorbed NDMA needs to be in contact with an eluent to be extracted. If water remains on the surface of the adsorbent, the eluent will not readily contact NDMA. This is because DCM is not readily mixed with water, which may lead to the incomplete extraction of NDMA. To improve the polarity of the eluent, methanol (5%) was added. However, this was not effective; rather, it decreased extraction efficiency possibly because the addition of methanol decreased the solubility of NDMA.

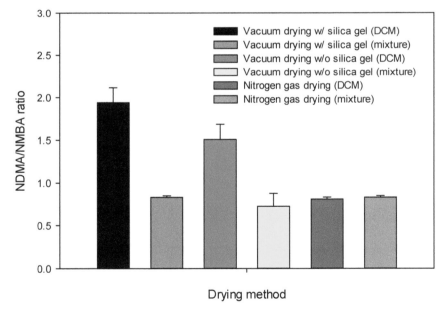

**Figure 2.** Comparison of the *N*-nitrosodimethylamine (NDMA) ratios with NMBA (a surrogate) peak areas among six different combinations of drying methods and eluent options (*n* = 3).

Drying with a nitrogen gas stream (fifth and sixth sets) resulted in area response ratios approximately half that of vacuum drying with a silica gel trap using DCM as an eluent and ratios similar to the second and fourth methods. This suggests a loss of NDMA during drying. Other flow rates and drying durations using a nitrogen stream were attempted, but there was no improvement.

Some previous studies used a nitrogen gas flow for adsorbent drying but did not report information regarding the gas flow rate or pressure [14,21].

Therefore, it is not easy to obtain reproducible experimental results without optimizing the two parameters. For instance, McDonald et al. [21] found a relatively low percent recovery of approximately 50% and 79% by using ultrapure and tap water as sample matrices at 10 and 100 ng/L levels, respectively. This suggests that in the eluate, NDMA might be incompletely extracted from the adsorbent because of incomplete drying, or it might be lost prior to sample elution because of excessive drying.

Of the three drying durations, 10 min was found to be optimal (Figure 3). No significant difference between 10 and 30 min was observed, and longer drying periods (60 and 90 min) led to a decrease in peak area ratios, likely because of NDMA loss as a result of the air stream purge. Furthermore, if the drying time is too long, the additional formation of NDMA on the adsorbent can occur because of the reaction of DMA with $N_2O$, nitrosamine ($H_2N_2O$), or $NH_2OH$ [23,24]. Among the activated carbons tested, Carboxen 572, which was adopted in this study, exhibited the least NDMA formation. Therefore, care needs to be taken to produce reliable measurement results during the sample preparation stage.

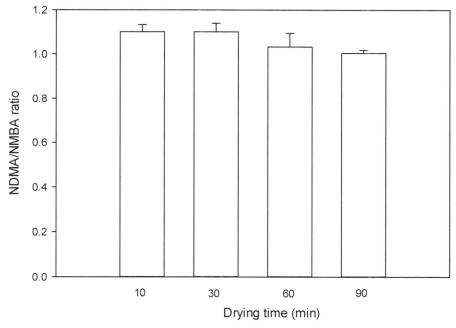

**Figure 3.** Optimization of drying duration for vacuum drying with a silica gel trap ($n = 3$).

The removal of DMA from water by washing the eluate with dilute HCl solution was evaluated, and the result is illustrated in Figure 4. Approximately 70% of DMA remained in the final analytical sample without the acid washing step. To the contrary, approximately 65% of DMA was removed after washing the 15 mL DCM eluate with a 1 N HCl solution, whereas complete removal was attained by washing the 1 mL DCM concentrate with a 1 N HCl solution. This finding suggests that acid washing is necessary for DMA removal, and the most efficient removal can be achieved by washing a small volume of the concentrate with the acid.

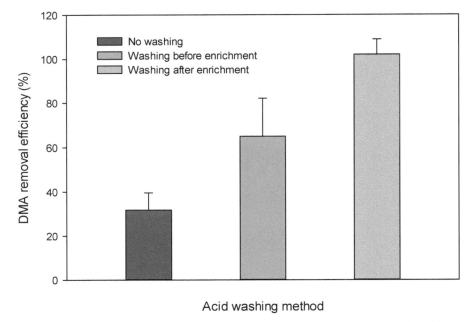

**Figure 4.** Comparison of the removal efficiency of dimethylamine (DMA) between two acid washing options. No acid washing resulted in a DMA removal of only 31.8 ± 7.8%. Washing the 15 mL eluate with a 1 N HCl solution prior to concentrating with a nitrogen gas stream resulted in the removal of 65.0 ± 17.1% of DMA. Concentrating the eluate to 1 mL and then washing it with 1 N HCl resulted in the complete removal (102.2 ± 6.9%) of DMA, indicating no overestimation of NDMA concentration because of the presence of DMA in water.

The final established method based on the above procedures is as follows. A 500 mL water sample was passed through a preconditioned cartridge (20 mL each of hexane, DCM, and methanol) containing 2.0 g of Carboxen® 572 (20–45 mesh) at a rate of 10 mL/min. The upper end of the cartridge was connected to a trap containing ~25 mL (~16 g) silica gel, and vacuum (−30 kPa) was applied for 10 min to dry the adsorbent. DCM was passed through the cartridge at 10 mL/min to elute NDMA and NMBA. The eluate was concentrated to 1 mL by a mild stream of nitrogen gas. One microliter of the 1 N HCl solution was added to the concentrate, and the mixture was vigorously shaken for 1 min using the vortex mixer. The bottom DCM layer was transferred to a 4 mL vial, to which ~50 mg of $Na_2SO_4$ was added. After the filtration of the mixture through a pipette packed with a small amount of defatted cotton, the denitrosating reagent (150 μL) was added, and the mixture was heated at 40 °C for 30 min. The denitrosating reagent and DCM were evaporated using a nitrogen gas stream, and a buffer solution (pH 10.5, 150 μL) and a dansylating reagent (150 μL) were then added to the residue. The mixture was heated at 40 °C for 30 min, and 50 μL of ultrapure water was added. Thereafter, the aliquots of 40 μL were injected for HPLC analysis.

*3.2. Results of Method Validation*

Table 2 presents the results of the method detection limit, limit of quantitation, linearity, accuracy (percent recovery and percent error), and precision (repeatability). The MDL was estimated to be 0.26 ng/L, which is comparable to those from previous studies (0.45 ng/L [21] and 0.28 ng/L [25]), in which GC-tandem MS was used for instrumental analysis. The calibration curve was linear with a coefficient of determination ($r^2$) of 0.9984. Percent recoveries and errors ranged from 81.1% to 92.1% and from 2.13% to 12.0%, respectively. The percent recoveries are similar to those of the EPA method

521 (83.7 to 94.7% for reagent water and drinking water fortified with NDMA) [25]. Relative standard deviations calculated at three levels were lower than 7.1%, which were also similar to those of the EPA method 521 (4.4 to 12%) [25]. The above results indicate that the current analytical method was validated for the analysis of NDMA in field water samples.

**Table 2.** Method validation results for the determination of NDMA in water.

| Parameter | | Working Concentration (ng/L) Tested | Value |
|---|---|---|---|
| Method Detection Limit | | | 0.26 ng/L |
| Limit of Quantitation | | | 0.82 ng/L |
| Linearity ($r^2$) | | 2.0–80 | 0.9984 |
| Accuracy | Percent Recovery | 2.0 | 89.5% |
| | | 30 | 81.1% |
| | | 60 | 92.1% |
| | Percent Error | 2.0 | 12.0% |
| | | 30 | 11.3% |
| | | 60 | 2.13% |
| Precision | Repeatability (RSD) | 2.0 | 7.09% |
| | | 30 | 3.98% |
| | | 60 | 2.61% |

### 3.3. Nationwide Distributions of NDMA Concentrations in Tap Water

NDMA was found in most samples with detection rates of 95.1% (39 of 41) and 100% in summer and winter, respectively. The overall average (±standard deviation) NDMA concentration was 46.6 (±22.7) ng/L, in the range of 0.13–80.7 ng/L (Table 3).

**Table 3.** Regional and seasonal distributions of NDMA concentrations (ng/L) in nationwide tap water samples.

| Sampling Region | Summer | | Winter | | Both Seasons | |
|---|---|---|---|---|---|---|
| | *n* | Concentration | *n* | Concentration | *n* | Concentration |
| Seoul | 5 | 21.8 ± 12.8 (0.13–32.9) | 5 | 70.2 ± 8.3 (60.2–77.8) | 10 | 46.0 ± 27.5 (0.13–77.8) |
| Gyeonggi | 5 | 46.9 ± 23.5 (18.0–75.4) | 5 | 46.2 ± 25.8 (12.7–70.8) | 10 | 46.6 ± 23.3 (12.7–75.4) |
| Gangwon | 5 | 20.7 ± 7.8 (11.3–29.1) | 5 | 52.5 ± 24.7 (19.3–79.3) | 10 | 36.6 ± 24.1 (11.3–79.3) |
| Chungcheong | 8 | 30.5 ± 23.7 (3.00–66.2) | 8 | 62.7 ± 12.7 (46.7–80.0) | 16 | 46.6 ± 24.8 (3.00–80.0) |
| Jeolla | 7 | 56.3 ± 9.4 (42.8–70.2) | 7 | 63.7 ± 8.4 (54.1–80.7) | 14 | 60.0 ± 9.4 (42.8–80.7) |
| Gyeongsang | 9 | 36.9 ± 16.7 (11.9–58.6) | 9 | 54.9 ± 21.4 (17.2–78.7) | 18 | 45.9 ± 20.8 (11.9–78.7) |
| Jeju | 2 | 2.49 ± 3.30 (0.13–4.86) | 2 | 8.02 ± 5.27 (4.29–11.7) | 4 | 5.26 ± 4.81 (0.13–11.7) |
| Total | 41 | 36.5 ± 20.5 (0.13–75.4) | 41 | 56.2 ± 20.9 (4.29–80.7) | 82 | 46.6 ± 22.7 (0.13–80.7) |

Previous studies demonstrated that NDMA concentrations varied significantly depending on the study region, sampling location (plant effluent/distribution system), season, source water (groundwater/surface water), and disinfectant (chlorine/chloramines), ranging from <MDL to 630 ng/L [1,7,11–14,33–36]. NDMA levels up to 189 ng/L were reportedly found in China [7] but were mostly within 50 ng/L [35,36]. These levels are similar to those observed in this study.

Most of the NDMA concentrations measured in this study exceeded the drinking water concentration (7 ng/L) corresponding to the $10^{-5}$ cancer risk from oral exposure estimated by the EPA IRIS [3]. Thirty-seven (90.2%) and 40 (97.6%) samples exceeded this value in summer and winter, respectively. Moreover, two samples (4.88%) in summer and 13 samples (31.7%) in winter had concentrations greater than 70 ng/L ($10^{-4}$ risk level). Therefore, residents of the Korean Peninsula might undergo excessive cancer risks from the ingestion of chlorinated tap water with/without boiling because NDMA is thermally stable, and its concentration would increase after a long boiling period because of decreased water volume [37].

The NDMA concentrations measured for tap water in this study (<0.26–80.7 ng/L) were much higher than those measured for finished water (<0.50–13 ng/L) at drinking water treatment plants in Korea from 2013 to 2015 [13]. This is probably because NDMA formation continues to occur throughout the distribution pipes after the water is disinfected in the treatment plants [14,38]. Average (±standard deviation) distances from each treatment plant to each sampling site were estimated to be 13.2 (±14.1) km with a range of 0.60 to 66.5 km, leading to various NDMA formations.

*3.4. Regional Variations of NDMA Concentrations*

Table 3 presents NDMA concentrations in the seven regions. The concentrations in the Korean Peninsula for the entire data set ranged from <0.13 (MDL) to 80.7 ng/L, whereas those in the Jeju region were much lower, ranging from <0.13 to 11.7 ng/L. Table 4 shows the regional and seasonal distributions of water quality parameters in nationwide tap water samples. The lower NDMA levels in the Jeju region could be attributed to the use of spring water as a source of drinking water, which might contain lower levels of organic precursors to NDMA (Table 4).

The one-way ANOVA conducted to determine the differences in mean NDMA concentrations among all the data of the seven areas indicated a significant difference ($p = 0.024$) at the 5% significance level. A subsequent post hoc analysis using Tukey HSD indicated a regional difference in the mean NDMA concentration only between the Jeju (the lowest concentration) and the Jeolla (the highest concentration) areas. Significant differences among the other six regions were not found. Excluding the Jeju data, the ANOVA was conducted separately for summer and winter data. A significant difference in the mean NDMA concentrations for the summer data was observed between the Seoul (21.8 ng/L) and Jeolla (56.3 ng/L) regions ($p = 0.019$; Tukey HSD). However, such a difference was not found for the winter data. The above results indicate that average NDMA concentrations in the Korean Peninsula were observed at relatively high levels, but concentrations did not significantly differ among the sampling sites.

Table 4. Regional and seasonal distributions of water quality parameters in nationwide tap water samples.

| Sampling Region | Season | DMA (µg/L) | Free Cl (mg/L) | Combined Cl (mg/L) | pH | Water Temp. (°C) | DOC (mg/L) | SUVA (L/mg-m) | Total Nitrogen (mg/L) | Nitrate (mg/L) |
|---|---|---|---|---|---|---|---|---|---|---|
| Seoul | S * | 0.57 ± 0.11 | 0.90 ± 0.44 | 0.04 ± 0.05 | 6.44 ± 0.06 | 29.1 ± 0.9 | 1.16 ± 0.65 | 1.37 ± 0.29 | 8.53 ± 1.82 | 7.27 ± 0.32 |
| | W ** | 0.63 ± 0.19 | 0.39 ± 0.18 | 0.18 ± 0.03 | 6.94 ± 0.04 | 9.42 ± 2.08 | 1.43 ± 0.37 | 0.78 ± 0.35 | 9.61 ± 3.56 | 10.7 ± 0.88 |
| Gyeong-gi | S | 0.81 ± 0.22 | 0.80 ± 0.28 | 0.05 ± 0.04 | 6.81 ± 0.17 | 28.1 ± 2.8 | 1.21 ± 0.19 | 1.59 ± 0.23 | 8.15 ± 0.48 | 7.11 ± 0.94 |
| | W | 0.93 ± 0.40 | 0.69 ± 0.30 | 0.12 ± 0.11 | 7.05 ± 0.04 | 12.0 ± 2.9 | 1.92 ± 0.22 | 0.94 ± 0.31 | 10.8 ± 0.8 | 9.21 ± 0.97 |
| Gangwon | S | 0.92 ± 0.19 | 0.81 ± 0.17 | 0.04 ± 0.03 | 6.86 ± 0.44 | 24.5 ± 2.8 | 1.49 ± 0.61 | 2.10 ± 1.76 | 8.04 ± 2.93 | 6.67 ± 4.21 |
| | W | 0.45 ± 0.10 | 0.69 ± 0.10 | 0.04 ± 0.02 | 7.07 ± 0.27 | 10.6 ± 1.6 | 1.31 ± 0.63 | 1.50 ± 0.47 | 7.28 ± 1.64 | 5.46 ± 2.38 |
| Chung-cheong | S | 0.94 ± 0.37 | 0.68 ± 0.36 | 0.05 ± 0.04 | 6.95 ± 0.46 | 28.9 ± 1.4 | 1.68 ± 1.03 | 1.57 ± 0.29 | 7.30 ± 1.98 | 6.02 ± 2.53 |
| | W | 0.86 ± 0.22 | 0.46 ± 0.32 | 0.14 ± 0.11 | 7.00 ± 0.27 | 11.5 ± 3.1 | 1.39 ± 0.49 | 1.22 ± 0.61 | 8.41 ± 3.65 | 7.14 ± 4.68 |
| Jeolla | S | 0.86 ± 0.18 | 0.90 ± 0.38 | 0.02 ± 0.01 | 6.57 ± 1.82 | 26.0 ± 1.8 | 1.07 ± 0.44 | 1.90 ± 0.57 | 3.73 ± 1.96 | 3.75 ± 2.87 |
| | W | 0.85 ± 0.38 | 0.48 ± 0.16 | 0.07 ± 0.05 | 6.95 ± 2.15 | 10.5 ± 2.6 | 1.20 ± 0.33 | 1.43 ± 0.58 | 5.96 ± 3.45 | 3.90 ± 2.65 |
| Gyeong-sang | S | 0.89 ± 0.66 | 0.92 ± 0.23 | 0.04 ± 0.02 | 6.35 ± 0.35 | 28.1 ± 2.2 | 1.11 ± 0.36 | 1.58 ± 0.52 | 3.25 ± 1.32 | 3.43 ± 2.92 |
| | W | 0.79 ± 0.39 | 0.45 ± 0.31 | 0.10 ± 0.03 | 7.56 ± 0.25 | 8.62 ± 2.40 | 1.83 ± 0.65 | 1.97 ± 2.69 | 5.96 ± 3.45 | 8.21 ± 4.76 |
| Jeju | S | 0.45 ± 0.01 | - | - | - | - | - | - | - | 7.42 ± 0.93 |
| | W | 0.57 ± 0.15 | - | - | - | - | - | - | - | 7.35 ± 0.79 |

*: Summer. **: Winter. -: The data are missing.

47

*3.5. Seasonal Variations of NDMA Concentrations*

Nationwide water samples were grouped into two seasons (summer and winter) and seven locations (Seoul, Gyeonggi, Gangwon, Chungcheong, Jeolla, Gyeongsang, and Jeju) (Table 3). The average NDMA concentrations in summer and winter were 36.5 and 56.2 ng/L, respectively. The Jeolla and Seoul regions had the highest concentrations of 56.3 and 70.2 ng/L in summer and winter, respectively, whereas Jeju recorded the lowest average concentrations of 2.49 and 8.02 ng/L in summer and winter, respectively. The average NDMA concentrations were higher in winter than in summer, except for the Gyeonggi region, where the concentrations were similar in both seasons (46.9 in summer vs. 46.2 ng/L in winter).

The paired *t*-test conducted for the seasonal concentration comparison of all the samples indicated significantly greater concentrations in winter than in summer with $p = 0.000$. This result indicates that NDMA formation is favored in colder months as opposed to warmer ones. The aforementioned outcome is in agreement with those of previous studies in which the concentrations of NAs, including NDMA, in drinking water samples in colder seasons (winter and fall) were higher than those in summer [11,18,38,39].

Higher NDMA concentrations in colder seasons can be explained by lower microbiological activity and lower photodegradation because of less sunlight [38,40]. Since the microbial activities of *Nitrosomonas* and *Nitrobacter* are very low in colder seasons, ammoniacal nitrogen is not readily oxidized to more highly oxidized forms, such as nitrite and nitrate [40,41]. Therefore, higher levels of ammoniacal nitrogen in winter than those in summer might cause chloramines to form at a higher rate [42], followed by the ready formation of NDMA via the pathway involving unsymmetrical dimethylhydrazine chloride (UDMH-Cl) [43].

*3.6. Relationship between NDMA Concentrations and Water Quality Parameters*

A multiple regression analysis was conducted for the entire data set (with the exception of Jeju) by setting the NDMA concentration as the dependent variable and all other water quality parameters (including DMA concentration, free residual chlorine, combined residual chlorine, total residual chlorine, pH, water temperature, DOC, SUVA$_{254}$, nitrate concentration, and total nitrogen concentration) as independent variables. Only a single parameter, water temperature, was included in the regression equation with $p = 0.000$ and $R^2 = 0.271$ (adjusted $R^2 = 0.261$) (Figure 5). A relatively small difference between $R^2$ and adjusted $R^2$ indicates that the following equation is a good description of NDMA formation:

$$NDMA = 71.7 - 1.30 \times \text{water temp}$$

Additionally, there was no problem regarding multicollinearity since variance inflation factor (VIF) values were below 10 (1.000–1.352) for all dependent variables.

The above equation indicates the favorable formation of NDMA at low water temperatures. However, other parameters were not significantly related to NDMA concentration ($p > 0.05$). This is in good agreement with previous studies, in which high NDMA formation was observed in colder seasons [11,18,38,39].

Multiple regression analyses were also conducted separately for six different regions except Jeju. The following significant regression equation was obtained only for the Jeolla data:

$$NDMA = 13.7 + 55.4 \times DMA + 5.84 \times SUVA + 80.7 \times \text{Combined Cl}$$

The coefficient of determination ($R^2$) for this regression equation was high at 0.936 (adjusted $R^2 = 0.917$), and the *p*-values for DMA, SUVA, and combined Cl were 0.000, 0.001, and 0.001, respectively. This is consistent with previous reports: NDMA formation increases as the DMA concentration, NH$_2$Cl concentration [44], and SUVA [45] increase.

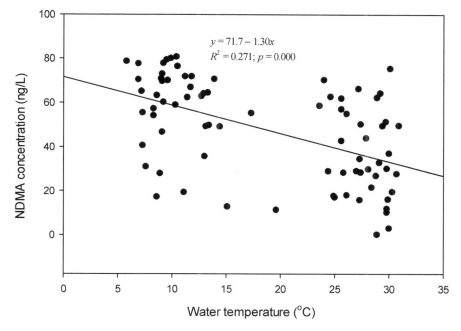

**Figure 5.** The relationship between water temperature and NDMA concentration. The two variables were significantly but weakly related to each other with $R^2 = 0.271$ and $p = 0.000$.

Some water treatment plants employ AOPs, including ozone treatment and PAC, as mentioned in Section 2.4. The variations in treatment processes could be one of the reasons for the weak relationship between NDMA concentration and other field water parameters. Therefore, excluding AOP data, multiple regression indicated a slightly increased $R^2$ (0.312) and adjusted $R^2$ (0.285) with two independent variables, such as water temperature and total chlorine concentration ($n = 56$). The following regression model is suggested for the data:

$$NDMA = 61.8 - 1.43 \times water\ temp + 17.4 \times T\text{-}Cl$$

Only a few of the parameters considered above were found to be significantly related to NDMA formation, suggesting that those factors complicatedly act on the reaction. Moreover, other parameters in addition to those listed above could be involved in the formation of NDMA. In the future, elegantly designed laboratory studies are necessary to determine such parameters.

The NDMA concentration is expected to decrease in the future in the authors' country because NDMA in treatment plant water should be quarterly monitored since 2018, and livestock raising is gradually modernized to minimize the discharge of animal wastes, one of its major organic precursor.

## 4. Conclusions

An analytical method using HPLC-FLD for the quantitative determination of NDMA in drinking water was optimized for SPE cartridge drying methods and acid washing for the removal of DMA from water. Vacuum drying through silica gel and acid-washing of the enriched eluate produced satisfactory method validation results.

NDMA concentrations were significantly higher in winter than in summer, suggesting that in colder seasons, NDMA formation is favored, whereas NDMA degradation occurs less favorably. This observation was confirmed by the multiple regression analysis of the entire data set, where water temperature was an important parameter included in the regression equation. NDMA concentrations varied significantly depending on sampling sites (except for the Jeju region), and they were distributed at high levels with an average of 47 ng/L and a maximum of 81 ng/L. Therefore, the oral consumption of NDMA in drinking water may pose excessive cancer risks for the residents of the Korean Peninsula.

This study is not without its limitations. The sample size was not large enough to observe statistical significance for the tests conducted, although samples were obtained from sites nationwide. Moreover, samples were only obtained from faucets. If a study on the NDMA concentrations throughout the distribution systems is conducted, more useful information regarding its formation could be obtained. Furthermore, studies to elucidate the sources of the high levels of NDMA and how its formation can be controlled are necessary to prevent possible health risks in these regions. Because relatively hydrophilic NDMA is not readily removed by the AOP, one of the favored method would be to reduce its formation by removing its organic precursor, DMA. This research is ongoing in the authors' laboratory, and the result will be published shortly.

**Author Contributions:** Conceptualization, H.K.; methodology, H.K., S.P. and S.J.; software, H.K. and S.P.; validation, S.P.; formal analysis, S.P. and S.J.; investigation, H.K.; resources, H.K., S.P. and S.J.; data curation, S.P. and H.K.; writing—original draft preparation, S.P.; writing—review and editing, H.K.; visualization, S.P.; supervision, H.K.; project administration, H.K.; funding acquisition, H.K.

**Funding:** This work was supported by the National Research Foundation of Korea (NRF) grant from the Korea government (MSIT) (No. 2015R1A2A203008216).

**Acknowledgments:** The authors would like to thank Enago (www.enago.co.kr) for the English language review. The authors are also grateful to assistant editor and reviewers for their insightful comments and suggestions.

**Conflicts of Interest:** The authors declare no conflict of interest.

## References

1. Krasner, S.W.; Mitch, W.A.; McCurry, D.L.; Hanigan, D.; Westerhoff, P. Formation, precursors, control, and occurrence of nitrosamines in drinking water: A review. *Water Res.* **2013**, *47*, 4433–4450. [CrossRef] [PubMed]
2. U.S. EPA. Monitoring Unregulated Drinking Water Contaminants. Available online: https://www.epa.gov/dwucmr/second-unregulated-contaminant-monitoring-rule#screening (accessed on 25 July 2019).
3. U.S. EPA. IRIS–N-Nitrosodimethylamine. Available online: https://cfpub.epa.gov/ncea/iris2/chemicalLanding.cfm?substance_nmbr=45 (accessed on 25 July 2019).
4. U.S. EPA. IRIS–Bromodichloromethane. Available online: https://cfpub.epa.gov/ncea/iris2/chemicalLanding.cfm?substance_nmbr=213 (accessed on 25 July 2019).
5. U.S. EPA. IRIS–Dichloroacetic Acid. Available online: https://cfpub.epa.gov/ncea/iris2/chemicalLanding.cfm?substance_nmbr=654 (accessed on 25 July 2019).
6. Han, K.; Kim, H. Optimization of analytical conditions for the determination of nitrosamines in chlorinated tap water by high performance liquid chromatography. *Anal. Sci. Technol.* **2010**, *23*, 551–559. (In Korean) [CrossRef]
7. Bei, E.; Shu, Y.; Li, S.; Liao, X.; Wang, J.; Zhang, X.; Chen, C.; Krasner, S. Occurrence of nitrosamines and their precursors in drinking water systems around mainland China. *Water Res.* **2016**, *98*, 168–175. [CrossRef] [PubMed]
8. EPA. IRIS–N-Nitrosodiethylamine. Available online: https://cfpub.epa.gov/ncea/iris2/chemicalLanding.cfm?substance_nmbr=42 (accessed on 25 July 2019).
9. Park, S.; Jung, S.; Kim, Y.; Kim, H. Determination of secondary aliphatic amines in surface and tap waters as benzenesulfonamide derivatives using GC-MS. *Anal. Sci. Technol.* **2018**, *31*, 96–105. (In Korean)
10. Asami, M.; Oya, M.; Kosaka, K. A nationwide survey of NDMA in raw and drinking water in Japan. *Sci. Total Environ.* **2009**, *407*, 3540–3545. [CrossRef] [PubMed]
11. Jurado-Sánchez, B.; Ballesteros, E.; Gallego, M. Occurrence of aromatic amines and N-nitrosamines in the different steps of a drinking water treatment plant. *Water Res.* **2012**, *46*, 4543–4555. [CrossRef] [PubMed]
12. Son, B.; Lee, L.; Yang, M.; Park, S.; Pyo, H.; Lee, W.; Park, J. Risk assessment and distribution characteristics of N-nitrosamines in drinking water treatment plants. *J. Korean Soc. Water Wastewater* **2018**, *32*, 389–398. (In Korean) [CrossRef]

13. Luo, Q.; Wang, D.; Wang, Z. Occurrences of nitrosamines in chlorinated and chloraminated drinking water in three representative cities. *China Sci. Total Environ.* **2012**, *437*, 219–225. [CrossRef]

14. Wang, W.; Ren, S.; Zhang, H.; Yu, J.; An, W.; Hu, J.; Yang, M. Occurrence of nine nitrosamines and secondary amines in source water and drinking water: Potential of secondary amines as nitrosamine precursors. *Water Res.* **2011**, *45*, 4930–4938. [CrossRef]

15. Charrois, J.W.A.; Boyd, J.M.; Froese, K.L.; Hrudey, S.E. Occurrence of *N*-nitrosamines in Alberta public drinking-water distribution systems. *J. Environ. Eng. Sci.* **2007**, *6*, 103–114. [CrossRef]

16. USEPA. UCMR 2 (2008–2010) Occurrence Data. 2012. Available online: https://www.epa.gov/dwucmr/occurrence-data-unregulated-contaminant-monitoring-rule#2 (accessed on 25 July 2019).

17. Zhao, Y.-Y.; Boyd, J.; Hrudey, S.E.; Li, X.-F. Characterization of new nitrosamines in drinking water using liquid chromatography tandem mass spectrometry. *Environ. Sci. Technol.* **2006**, *40*, 7636–7641. [CrossRef]

18. Woods, G.C.; Trenholm, R.A.; Hale, B.; Campbell, Z.; Dickenson, E.R.V. Seasonal and spatial variability of nitrosamines and their precursor sources at a large-scale urban drinking water system. *Sci. Total Environ.* **2015**, *520*, 120–126. [CrossRef] [PubMed]

19. Krasner, S.W.; Lee, C.F.T.; Mitch, W.A.; von Gunten, U. *Development of a Bench-Scale Test to Predict the Formation of Nitrosamines*; Water Research Foundation: Denver, CO, USA, 2012; pp. 1–98.

20. Pozzi, R.; Bocchini, P.; Pinelli, F.; Galletti, G.C. Determination of nitrosamines in water by gas chromatography/chemical ionization/selective ion trapping mass spectrometry. *J. Chromatogr. A* **2011**, *1218*, 1808–1814. [CrossRef] [PubMed]

21. McDonald, J.A.; Harden, N.B.; Nghiem, L.D.; Khan, S.J. Analysis of N-nitrosamines in water by isotope dilution gas chromatography–electron ionisation tandem mass spectrometry. *Talanta* **2012**, *99*, 146–154. [CrossRef] [PubMed]

22. Jung, S.; Kim, D.; Kim, H. Improving the chromatographic analysis of *N*-nitrosamines in drinking water by completely drying the solid sorbent using dry air. *Pol. J. Environ. Stud.* **2016**, *25*, 2689–2693. [CrossRef]

23. Padhye, L.; Hertzberg, B.; Yushin, G.; Huang, C.-H. *N*-Nitrosamines formation from secondary amines by nitrogen fixation on the surface of activated carbon. *Environ. Sci. Technol.* **2011**, *45*, 8368–8376. [CrossRef]

24. Padhye, L.; Wang, P.; Karanfil, T.; Huang, C.-H. Unexpected role of activated carbon in promoting transformation of secondary amines to N-nitrosamines. *Environ. Sci. Technol.* **2010**, *44*, 4161–4168. [CrossRef]

25. US EPA. *Method 521: Determination of Nitrosamines in Drinking Water by Solid Phase Extraction and Capillary Column Gas Chromatography with Large Volume Injection and Chemical Ionization Tandem Mass Spectrometry (MS/MS)*; National Exposure Research Laboratory, Office of Research and Development, US Environmental Protection Agency: Washington, DC, USA, 2004.

26. Qian, Y.; Wu, M.; Wang, W.; Chen, B.; Zheng, H.; Krasner, S.W.; Hrudey, S.E.; Li, X.-F. Determination of 14 nitrosamines at nanogram per liter levels in drinking water. *Anal. Chem.* **2015**, *87*, 1330–1336. [CrossRef]

27. Komarova, N.V.; Velikanov, A.A. Determination of volatile *N*-nitrosamines in food by high-performance liquid chromatography with fluorescence detection. *J. Anal. Chem.* **2001**, *56*, 359–363. [CrossRef]

28. Wang, Z.; Xu, H.; Fu, C. Sensitive fluorescence detection of some nitrosamines by precolumn derivatization with dansyl chloride and high-performance liquid chromatography. *J. Chromatogr. A* **1992**, *589*, 349–352. [CrossRef]

29. Cha, W.; Nalinakumari, B.; Fox, P. High-performance liquid chromatography for determination of *N*-nitrosodimethylamine in water. *Proc. Water Environ. Fed.* **2006**, *2016*, 889–900. [CrossRef]

30. Sacher, F.; Lenz, S.; Brauch, H.-J. Analysis of primary and secondary aliphatic amines in waste water and surface water by gas chromatography-mass spectrometry after derivatization with 2,4-dinitrofluorobenzene or benzenesulfonyl chloride. *J. Chromatogr. A* **1997**, *764*, 85–93. [CrossRef]

31. Ma, F.; Wan, Y.; Yuan, G.; Men, L.; Don, Z.; Hu, J. Occurrence and source of nitrosamines and secondary amines in groundwater and its adjacent Jialu River Basin, China. *Environ. Sci. Technol.* **2012**, *46*, 3236–3243. [CrossRef] [PubMed]

32. Ebina, J.; Tsutsui, T.; Shirai, T. Simultaneous determination of total nitrogen and total phosphorus in water using peroxodisulfate oxidation. *Water Res.* **1983**, *17*, 1721–1726. [CrossRef]

33. Krasner, S.W. The formation and control of emerging disinfection by-products of health concern. *Philos. Trans. R. Soc. A* **2009**, *367*, 4077–4095. [CrossRef] [PubMed]

34. Russell, C.G.; Blute, N.K.; Via, S.; Wu, X.; Chowdhury, Z.; More, R. Nationwide assessment of nitrosamine occurrence and trends. *J. Am. Water Work. Assoc.* **2012**, *104*, E205–E217. [CrossRef]

35. Wang, W.; Yu, J.; Ana, W.; Yang, M. Occurrence and profiling of multiple nitrosamines in source water and drinking water of China. *Sci. Total Environ.* **2016**, *551*, 489–495. [CrossRef]
36. Zhao, C.; Lu, Q.; Gu, Y.; Pan, E.; Sun, Z.; Zhang, H.; Zhou, J.; Du, Y.; Zhang, Y.; Feng, Y.; et al. Distribution of N-nitrosamines in drinking water and human urinary excretions in high incidence area of esophageal cancer in Huai'an, China. *Chemosphere* **2019**, *235*, 288–296. [CrossRef]
37. Kim, H.; Han, K. Ingestion exposure to nitrosamines in chlorinated drinking water. *Environ. Health Toxicol.* **2011**, *26*, e2011003. [CrossRef]
38. Zhang, A.; Li, Y.; Chen, L. Distribution and seasonal variation of estrogenic endocrine disrupting compounds, N-nitrosodimethylamine, and N-nitrosodimethylamine formation potential in the Huangpu River, China. *J. Environ. Sci.* **2014**, *26*, 1023–1033. [CrossRef]
39. Uzun, H.; Kim, D.; Karanfil, T. Seasonal and temporal patterns of NDMA formation potentials in surface waters. *Water Res.* **2015**, *69*, 162–172. [CrossRef] [PubMed]
40. Layton, A.C.; Gregory, B.W.; Seward, J.R.; Schultz, T.W.; Sayler, G.S. Mineralization of steroidal hormones by biosolids in wastewater treatment systems in Tennessee USA. *Environ. Sci. Technol.* **2000**, *34*, 3925–3931. [CrossRef]
41. NIER. *A Study on the Improvement of the Environmental Standard Methods for Pretreatment Process: Establishment of an Official Testing Method for Drinking Water Using Autoanalyzer*; National Institute of Environmental Research: Incheon, Korea, 2008.
42. Sudarno, U.; Winter, J.; Gallert, C. Effect of varying salinity, temperature, ammonia and nitrous acid concentrations on nitrification of saline wastewater in fixed-bed reactors. *Bioresour. Technol.* **2011**, *102*, 5665–5673. [CrossRef] [PubMed]
43. Jafvert, C.T.; Valentine, R.L. Reaction scheme for the chlorination of ammoniacal water. *Environ. Sci. Technol.* **1992**, *26*, 577–586. [CrossRef]
44. Choi, J.; Valentine, R.L. Formation of N-nitrosodimethylamine (NDMA) from reaction of monochloramine: A new disinfection by-product. *Water Res.* **2002**, *36*, 817–824. [CrossRef]
45. Chen, Z.; Valentine, R.L. Formation of N-nitrosodimethylamine (NDMA) from humic substances in natural water. *Environ. Sci. Technol.* **2007**, *41*, 6059–6065. [CrossRef]

*Article*

# Mapping Dynamics of Bacterial Communities in a Full-Scale Drinking Water Distribution System Using Flow Cytometry

Caroline Schleich [1], Sandy Chan [2,3,4], Kristjan Pullerits [2,3,4], Michael D. Besmer [5], Catherine J. Paul [2,6], Peter Rådström [2] and Alexander Keucken [1,6,*]

[1]  Vatten & Miljö i Väst AB, SE-311 22 Falkenberg, Sweden; Caroline.Schleich@vivab.info
[2]  Applied Microbiology, Department of Chemistry, Lund University, P.O. Box 124, SE-221 00 Lund, Sweden; Sandy.Chan@sydvatten.se (S.C.); kristjan.pullerits@tmb.lth.se (K.P.); catherine.paul@tvrl.lth.se (C.J.P.); peter.radstrom@tmb.lth.se (P.R.)
[3]  Sweden Water Research AB, Ideon Science Park, Scheelevägen 15, SE-223 70 Lund, Sweden
[4]  Sydvatten AB, Hyllie Stationstorg 21, SE-215 32 Malmö, Sweden
[5]  onCyt Microbiology AG, CH-8038 Zürich, Switzerland; michael.besmer@oncyt.com
[6]  Water Resources Engineering, Department of Building and Environmental Engineering, Faculty of Engineering, Lund University, P.O. Box 118, SE-221 00 Lund, Sweden
*  Correspondence: alexander.keucken@vivab.info

Received: 9 September 2019; Accepted: 11 October 2019; Published: 15 October 2019

**Abstract:** Microbial monitoring of drinking water is required to guarantee high quality water and to mitigate health hazards. Flow cytometry (FCM) is a fast and robust method that determines bacterial concentrations in liquids. In this study, FCM was applied to monitor the dynamics of the bacterial communities over one year in a full-scale drinking water distribution system (DWDS), following implementation of ultrafiltration (UF) combined with coagulation at the drinking water treatment plant (DWTP). Correlations between the environmental conditions in the DWDS and microbial regrowth were observed, including increases in total cell counts with increasing retention time (correlation coefficient $R = 0.89$) and increasing water temperature (up to 5.24-fold increase in cell counts during summer). Temporal and spatial biofilm dynamics affecting the water within the DWDS were also observed, such as changes in the percentage of high nucleic acid bacteria with increasing retention time (correlation coefficient $R = -0.79$). FCM baselines were defined for specific areas in the DWDS to support future management strategies in this DWDS, including a gradual reduction of chloramine.

**Keywords:** flow cytometry; biofilm; drinking water distribution system; ultrafiltration; coagulation; drinking water management

## 1. Introduction

Drinking water needs to be safe, esthetically acceptable, and not cause excessive damage to infrastructure. These aspects of water quality are impacted by microorganisms, the majority of which are bacteria that originate from the source water, are shaped by processes in the drinking water treatment plant (DWTP), and are contributed from biofilms in the drinking water distribution system (DWDS) during distribution [1–4]. A high bacterial cell concentration can lead to: Esthetic problems, such as discoloration of the water and/or changes in taste and odor; increased biocorrosion with concomitant high copper and iron concentrations in the water; and thus deterioration of the DWDS [5–7]. Growth of opportunistic pathogens such as *Legionella ssp.* in the drinking water can pose a severe health risk [8,9]. To counter these risks, the DWTP should control bacterial survival and regrowth in the DWDS, using methods like filtration, which limits the input of nutrients, and disinfection, using UV irradiation and chlorination [4,10,11]. Some bacteria, however, often remain in the drinking water after these and other treatments, and enter the DWDS [12]. The estimated bacterial concentration in most distributed drinking water is between $10^6$ to $10^8$ cells/L [13,14]. While these high bacterial counts are generally considered to have no direct impact on public health [9], abrupt changes in bacterial concentrations can indicate failure of disinfection or filtration, or pipe breakage, that could indicate occurrences in the treatment process or external contamination in the DWDS that could indirectly impact the consumer [7]. Detecting any sudden changes, however, requires an understanding of which bacterial counts are expected, with this knowledge generated by comprehensive monitoring of the bacterial community in a DWDS.

Conventional bacterial monitoring of process performance is largely based on enumeration of indicator bacteria, such as *Escherichia coli*, coliforms and heterotrophs in grab samples of water [15]. These methods are labor- and resource-intensive and thus expensive. In addition, these methods detect only specific fractions of the bacterial community [16], limiting their resolution for detailed studies of bacterial regrowth in a DWDS [7]. Flow cytometry (FCM) has been proposed as a modern, rapid, standardized, and increasingly used alternative detection method for bacteria in drinking water [13,17]. This laser-based method rapidly, accurately, and reproducibly determines the concentration of bacteria in a water sample [18,19], and can also be used to measure the number of intact cells within the total population to assess the effectiveness of some treatments, such as chlorination [20]. Changes in the type of bacteria within the community are assessed by observing fluctuations in the distribution of DNA within the cells; for example, by comparing the distribution of cells across populations defined by the user (gates), such as high nucleic acid (HNA) bacteria and low nucleic acid (LNA) bacteria [14].

While there is broad consensus regarding typical values from source to tap for FCM-based bacterial concentrations in drinking water, specific baselines and the range of fluctuations around these baselines that are consistent with safe water need to be established for each DWDS individually. This requires large data sets collected from the drinking water treatment and distribution systems of interest. These need to define routine values, to describe proper functioning of the whole system, and identify how, and to what degree, fluctuations in these values reflect abnormalities and can describe the success of corrective actions. Baselines generated by permanent surveillance of routine operations may also be valuable for planning and monitoring changes in the treatment or distribution of water.

In this study, the process of establishing FCM baselines for a drinking water treatment and distribution system is described. Historically, Kvarnagården DWTP in Varberg, Sweden had limited treatment of surface water and very high cell concentrations ($7 \times 10^5$ cells/mL) of bacteria in the DWDS. In November 2016, the DWTP was upgraded to ultrafiltration (UF) membranes, combined with flocculation, reducing the input of bacterial cells into the DWDS and removing about 50% of natural, especially high molecular weight, organic carbon [2,21]. This upgrade was closely monitored with FCM and showed that bacterial concentrations in treated and distributed water were substantially lowered by the change in treatment processes [2]. With extremely low numbers of bacteria contributed by the treatment plant, the contribution of cells from biofilms in the DWDS to the total bacterial concentrations was observed, and the low bacterial concentrations at several monitoring locations in the DWDS indicated that concerns about an initial massive detachment of biofilm due to the changes in treatment were unwarranted [1,2].

To further monitor the biofilm over a longer time period, and to expand the application of FCM monitoring in this DWDS, additional sampling locations at greater distances and incorporation of different hydrodynamic and material properties in the DWDS were examined. The extensive sampling campaign of the selected sampling points was conducted over 12 months.

The objectives of this study were to:
(1) Confirm that the biofilm did not detach in the long-term.
(2) Assess the impact of seasonal changes on cell concentrations.
(3) Obtain detailed, spatially resolved information throughout the DWDS.
(4) Gather insights on driving environmental and/or technical factors of cell concentrations.

## 2. Materials and Methods

### 2.1. Study Site and Sampling

The study location, a DWDS in Varberg, Sweden operated by the utility VIVAB, is comprised of roughly 580 km of pipes, of which the majority are polyvinyl chloride (35%) and polyethylene (20%). This DWDS distributes approximately 5 million m$^3$ water annually to 60,000 residents, and is produced from surface water by the DWTP Kvarnagården. The treatment process consists of rapid sand filtration, UF combined with a coagulation step, pH adjustment, and disinfection with UV and chloramine (between 0.13 and 0.21 mg/L; Supplementary Materials Figure S1). Eighteen different locations were sampled from April 2018 to April 2019, beginning approximately one and a half years after the installation of the UF membrane at the DWTP in November 2016. This included the UF membrane treatment process (feed water, permeate) and the outgoing water from the DWTP with 15 points located in the DWDS (Supplementary Materials Figure S2). In total, 510 samples were taken and analyzed at two-week sampling intervals, with weekly sampling during July and September and no sampling during August. All water samples were collected in sterile 15 mL Falcon tubes with the addition of 1% (*v/v*) sodium thiosulphate (20 g/L) for quenching residual chlorine. The sampling routine included burning off the tap and flushing the line for 10 min before sampling and recording of water temperature. Samples were transported in cooling boxes and analyzed by FCM the same day.

### 2.2. FCM Analysis

FCM analysis was performed on a BD Accuri C6 Flow Cytometer (BD Biosciences, Belgium) with a 50 mW argon laser, wavelength = 488 nm [22]. Fluorescence from SYBR® Green I (Invitrogen AG, Switzerland) and propidium iodide were read at 533 ± 30 nm = FL1 (green fluorescence) and > 670 nm = FL3 (red fluorescence), respectively. The flow rate was 35 μL/min with a threshold of 500 arbitrary units of green fluorescence. Samples were stained with 5 μL of SYBR Green I at 100× diluted with dimethyl sulphoxide in a total volume of 500 μL corresponding to 1 × SYBR Green I final concentration and incubated in the dark for 15 min at +37 °C. When included, the concentration of propidium iodide was 0.3 mM (Sigma-Aldrich, Germany). Identical gates were applied for both types of staining (intact cell count (ICC) and total cell count (TCC)).

### 2.3. Other Water Quality Parameters

Total chlorine was analyzed using a SL1000 Portable Parallel Analyzer (Hach, Düsseldorf, Germany). Water temperature was measured with a TD 10 Thermometer (VWR, Radnor, PA, United States). TOCeq was measured online using a s::can (i::scan™; s::can Messtechnik GmbH, Vienna, Austria) with wavelength range 230–350 nm. The online absorbance measurements were calibrated against laboratory TOC analyses from the laboratory to calculate TOCeq.

### 2.4. Data Analysis

For FCM, manual gating strategies and a pattern analysis approach were applied. FCM fingerprints, including ratios of LNA and HNA bacteria, were compared and analyzed using the single cell analysis software FlowJo (Treestar, Inc., San Carlos, CA, USA). CHIC analysis (CHIC: Cytometric Histogram Image Comparison) was used for pattern analyses using R packages flowCHIC and flowCore [23–25]. FCM scatterplots were converted into 300 × 300 pixel images with 64-channel gray scale resolution for image comparison. The values generated by CHIC describing the differences between water samples were visualized using a non-metric multidimensional scale (NMDS) plot based on Bray–Curtis dissimilarity to capture changes in bacterial community structure [23,24]. To simplify correlations, the *envfit* function for environmental vectors was applied using R software [26].

The contact area between water and biofilm was determined by calculating the ratio between lateral surface and water volume for each pipe segment (Equation (1)).

$$\text{Contact area} = \frac{2\Pi r h}{\Pi r 2\, h} \tag{1}$$

Hydraulic modeling using MIKE Urban (DHI) was applied to determine specific retention times for different pressure zones in the DWDS.

## 3. Results and Discussion

### 3.1. Long-Term Stability After an Upgrade of the Treatment Process

The implementation of UF combined with coagulation at the DWTP Kvarnagården in Varberg led to a significant change in water quality [2,27]. TCC was reduced by a factor of $10^3$ cells/mL and about 50% of natural organic carbon, especially the high molecular weight fraction was removed by direct coagulation over the UF membranes (Figure 1). Before the changes in the treatment process, the number of bacteria in the water was approximately 660,000 ± 7000 cells/mL regardless of sampling site and time. After installation of the new process, TCC diminished to about 27,500 ± 9600 cells/mL at all sampling points, reaching a low of 3400 ± 2000 cells/mL in February 2017. TCC in the outgoing drinking water continued to decrease to about 350 ± 170 cells/mL in March 2018, giving a 1000-fold reduction of TCC in produced drinking water due to installation of UF. TCC determined for an expanded number of DWDS sampling points from April 2018 until April 2019 showed consistent and expected cell counts at all sampling points, subject to seasonal variations, and no large, rapid changes in TCC in the water phase that could indicate detachment of biofilm [20].

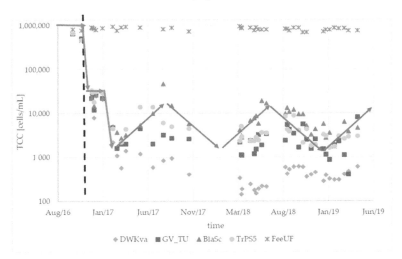

**Figure 1.** Total cell count (TCC) in water from the drinking water treatment plant (DWTP) and drinking water distribution system (DWDS). TCC was measured in the feed water to the ultrafiltration (UF) membrane (FeeUF, red stars); outgoing drinking water (DWKva, blue diamonds); and distributed water at an office building located at WWTP Getteröverket (GV_TU, purple squares), a public school in Bläshammar (BlaSc, green triangles), and a pump station in Trönningenäs (TrPS5, yellow circles). Measurements were taken before and after the installation of UF, indicated by the vertical dashed line. The red arrows show changes of TCC before, during, and shortly after commissioning of UF; blue arrows indicate seasonal changes in TCC. The graph includes published data (until January 2017, [2]).

### 3.2. Seasonal Changes in the Bacteral Community

The extended sampling program took place from April 2018 to April 2019 (Supplementary Materials Figure S2). Water temperatures increased at all sampling points during the summer, as expected, from 6.68 ± 1.28 °C in April 2018 to 13.68 ± 2.83 °C in September 2018, an increase of 6.99 ± 2.30 °C (Supplementary Materials Figure S3).

The average TCC during April 2018, considering all DWDS sampling points, was $3.1 \times 10^4$ ± $3.6 \times 10^4$ (range: $3 \times 10^2$ to $1.1 \times 10^5$ cells/mL). This increased in summer to an average TCC of $1.0 \times 10^5$ cells/mL (range: $5 \times 10^2$ to $4 \times 10^5$ cells/mL), an average 3.35-fold increase (range: 1.51–5.24-fold increase; Supplementary Materials Figure S3). This may be partially explained by regrowth of bacteria due to the elevated water temperatures during summer.

To demonstrate the observed seasonal shifts, values obtained from water collected at sampling point Masar showed TCC fluctuating with changes in water temperature with an increase of TCC during the summer months (Figure 2). The TCC in the outgoing water from the DWTP also increased during this period, from 230 ± 70 cells/mL in April to 540 ± 80 cells/mL in September (temperature increase from 5.8 to 7.5 °C). At Masar, the TCC was 24,400 ± 330 cells/mL in April with a water temperature of 6.3 °C. The water temperature at this sampling point increased to 14.9 °C and TCC increased by 4-fold to 98,300 ± 1200 cells/mL. By April 2019, TCC had returned to 23,350 ± 300 cells/mL with a water temperature of 6.2 °C, showing a clear seasonal trend. The ratio of intact bacteria in the water at Masar also fluctuated seasonally, from 69 ± 1% intact cells in April to 86 ± 2% in September, with an increase in intact cells supporting the observation that increases in TCC during these months are due to bacterial growth.

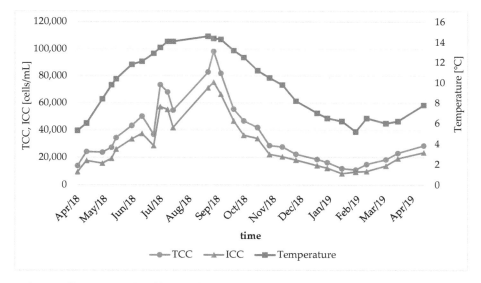

**Figure 2.** Changes in TCC (red line, circles), intact cell count (ICC) (green line, triangles), and water temperature (blue line; squares) at sampling point Masar. A clear seasonal trend, with increases in TCC and water temperature during the summer period, is shown.

The bacterial community, indicated by nucleic acid content described by FCM, also changed with season. The bacterial population gradually shifted to greater numbers of bacteria described as LNA during the summer months (Figure 3). Analyses from later months indicated a return of the bacterial community to a previous state (both in TCC and community structure, data not shown). A study from 2014 showed equal temporal trends in the microbial community that indicate annual reproducibility [28]. The histograms shown in Figure 3 represent individual FCM fingerprints of microbial communities. The amount of DNA in a single bacteria cell is described by its position along the x-axis. Cells registering with higher green fluorescence contain increasing amounts of DNA [22].

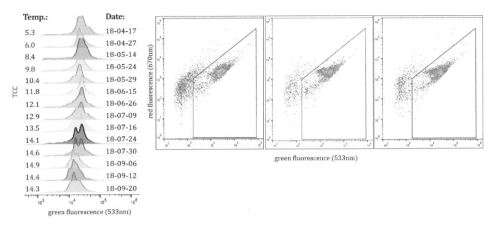

**Figure 3.** Changes in the bacterial community over season at Masar. FCM fingerprints (left panel) from sampling in April 2018 to September 2018 show a shift to increasing proportions of low nucleic acid (LNA) bacteria, likely due to an increase in water temperature during summer. Scatterplots (right panel) show individual cells within the population, the gate applied, and a gradual increase in intensity corresponding to more cells with LNA (sampling date for left to right: 27/04/18; 16/07/18; 24/07/18).

As expected, more bacterial growth (regrowth) was observed when the temperature in the water rose. Increasing TCC and temperature were also observed at Tofta, Tronn, and BlaSc, with 7000 ± 6200 cells/mL and water temperature: 7.2 ± 1.4 °C during spring, rising to 34,300 ± 28,000 cells/mL and water temperature: 15.5 ± 2.1 °C during summer (5.16 ± 0.89-fold increase in TCC; Supplementary Materials Figure S3). Seasonal changes for bacterial communities in DWDS have been observed in previous studies, and indicate reproducible annual patterns [28–30]. However, in the current study, some sampling points did not show this expected trend. Water from Hunst, Himle, Godst, and Lofta had a high TCC during spring (Supplementary Materials Figures S3 and S5) and an increase in water temperature had only a moderate impact on the TCC. TCC at these points increased from 55,400 ± 14,700 cells/mL (water temperature: 7.2 ± 0.3 °C) during spring to 156,200 ± 38,600 cells/mL (water temperature: 15.8 ± 0.8 °C) during summer, giving only a 2.84 ± 0.13-fold increase associated with the warmer water temperatures. Water from sampling point TrPS5, despite having an increase in water temperature of 7.1 °C (from 5.6 to 12.7 °C), increased TCC by only a factor of 1.79 (from 2900 ± 70 to 5200 ± 100 cells/mL; Supplementary Materials Figure S6). The apparent decoupling between temperature and TCC at this specific sampling location maybe be due to a combination of the low retention time and the short distance between the sampling point and the DWTP, which together would prevent significant accumulation of bacteria, either from regrowth or cells leaving the biofilm, to the water [2].

*3.3. Detection of Abnormalities Within the DWDS Using TCC Baselines*

Baselines were defined for each water sampling point in the study by grouping months with similar TCC values (standard deviation ≤25%) and then calculating average TCC values for those periods. Warning and alarm limits were determined to describe tailored acceptable TCC values for each sampling point and period. Applications of these defined baselines detected an abnormality within the DWDS in late June 2019. During routine sampling, the TCC at sampling point TrPS5 was 717,000 cells/mL, exceeding the alarm limit of 15,000 cells/mL for this location (Supplementary Materials Figure S7). Consultations with the personnel responsible for the DWDS revealed that maintenance work had occurred near TrPS5 in March 2019 and further investigation discovered that a valve near this sampling point had been accidentally closed. This resulted in insufficient circulation of the water, and possibly bacterial regrowth due to stagnation, and provided a likely explanation for the increased TCC [14]. The TCC returned to normal after the opening of the valve. While the most likely explanation for the increase in TCC is bacterial growth, additional explanations could include increased detachment of biofilm due to altered fluid dynamics in the pipe [31] or increased contact time of the stagnant water with the biofilm [32].

*3.4. Relationship of the Bacterial Population to Contact Time with Biofilm*

Possible correlations between TCC, %HNA, %ICC, and characteristics of the DWDS were evaluated using CHIC in combination with the *envfit* function (Figure 4). CHIC analysis compares individual FCM scatterplots describing the bacterial community in water samples. The comparisons are presented as a non-metric multidimensional scale (NMDS) plot, with water samples containing similar bacterial content shown as two dots that are close to each other, and water samples with more dissimilar bacterial content in the water placed further away from each other. This enables a visual determination of changes in the different bacterial communities, with the content of all water samples in one study presented in the same plot. The addition of vector analysis enables visualization of correlations between the bacterial communities, as determined by FCM, and environmental parameters. Longer vectors indicate stronger correlations, and those pointing in opposite directions indicate negative correlations.

Correlations between TCC and retention time, and TCC and the contact area between biofilm and the water, were determined using three data sets for each sampling point (Supplementary Materials Table S1). TCC increased with increasing retention time, as well and increasing contact area (Figure 4A). There was a negative correlation between retention time and %HNA bacteria. The amount of HNA bacteria in the water samples decreased with an increase in retention time. These results were confirmed by a more detailed examination using a subset of samples selected from the range of 0–50 in the NMDS in Figure 4A. This reduced the influence from outliers, such as samples with, for example, a high retention time (Figure 4B). When fewer samples were included in the analysis, the relationship between the correlations of %HNA and retention time shifted slightly (Figure 4B). However, more precise analyses are necessary to confirm these assumptions.

The sampling points are connected to the DWTP by 87.7 km pipes consisting of 6 different pipe materials with a total pipe surface area of 79,900 m$^2$. Water from sampling points at a greater distance from the DWTP usually had a higher TCC than water sampled closer to the DWTP (Supplementary Materials Figure S3). During April 2018, water leaving the DWTP contained 230 ± 70 cells/mL; water from close to the DWTP (GV_TU, BlaSc) contained 2250 ± 990 cells/mL and a more distant sampling point (Björk) had water containing 113,200 ± 4100 cells/mL. One explanation for the increased cell counts with respect to distance was the increased contact time between the water and the pipe surface biofilm [2]. The contact area between the water and biofilm was then determined by incorporating both pipe lengths and diameters to examine the relationship between TCC and the contact area/mL (Figure 5). This identified a positive correlation between the number of cells in the water and the contact area provided by the pipe biofilm (correlation coefficient $R = 0.75$; $p = 0.0008$).

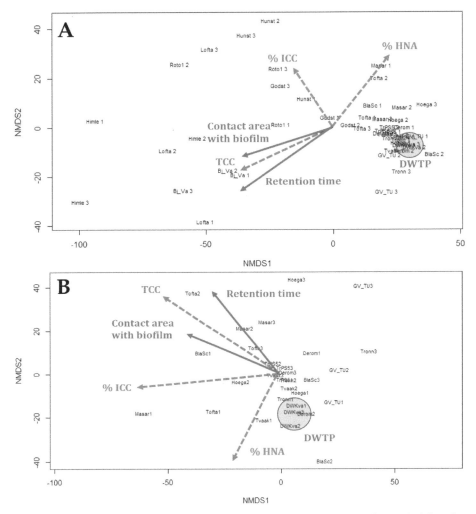

**Figure 4.** Output from CHIC analysis plot combined with *envfit* showing correlations (red dotted arrows: Bacterial community parameters; blue arrows: Distribution system parameters). Samples that are close to each other in the plot have similar bacterial profiles by FCM; vectors indicate possible correlations. (**A**) Analysis of 16 sampling points (*n* = 3 for each sampling point) sampled during April and May 2018. (**B**) Analysis of 10 sampling points, selected from the 16 in (**A**) (*n* = 3 for each sampling point sampled during April and May 2018. Using the smaller subset of samples permitted a more detailed examination of the correlations shown in panel (**A**) (NMDS1 range = 0–50).

Chlorine residues can influence microbial regrowth, which also changes with retention time and pipe length [29,33]. In this study, water leaving the DWTP contained on average 0.19 ± 0.02 mg/L chloramine; however, as expected, these chloramine concentrations were not present at all points in the distribution system (Supplementary Materials Figure S4). With the exception of a few sampling points close to the DWTP, the total chlorine concentration was below 0.04 mg/L at all other sampling points, and thus below the detection limit. A lack of chlorine can lead to faster regrowth of bacteria [33], and a free chlorine concentration of at least 0.3 mg/L is required to prevent bacterial regrowth in the distribution system [9]. The only impact that was observed was an inverse correlation between the intact cell concentration and chloramine concentration, with intact cell concentration elevated at sampling points where the total chlorine concentration was below 0.05 mg/L (Supplementary Materials Figure S4).

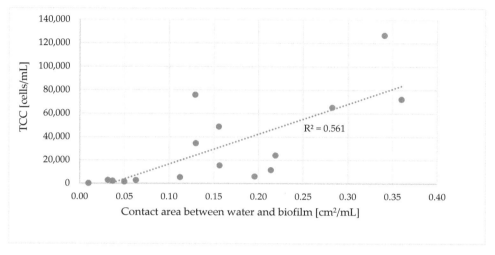

**Figure 5.** Correlation between the contact area for biofilm and water, and TCC.

This suggests that while changes in chlorine influenced whether the bacteria were intact or not, most of the bacterial dynamics between the water and biofilm observed in the current study as increases in TCC or changed %HNA are not due to loss of chlorine. Instead, a continuous interaction between the water and the biofilm, with the growth and release of bacteria from biofilm, provides a gradually accumulating supply of planktonic bacteria that can be monitored by FCM in the bulk water. A previous study calculated that it takes 2.31 days to double the number of planktonic bacteria in a bioreactor-based model system for DWDS at 13 °C without chlorine [34]. This can be extrapolated to suggest that for growth of the bacteria released from the biofilm, a retention time of at least 55 hours is required to double their number, and at shorter times, changes in the bacterial content of the water must be attributed to addition of cells from biofilm [2]. It should, however, be noted that monitoring of the biofilm by FCM was possible in this current study, and could be applied in other DWDS for inferring biofilm interactions, because the input of cells from the treatment plant is minimal [2]. When the number of cells in the outgoing drinking water is high, due to processes in the DWTP such as biological filtration, the resolution of FCM prevents detailed descriptions of bacteria only entering the water from DWDS biofilm or regrowth.

Prolonged stagnation of water has been associated with significant growth [34], and longer retention times also contribute to increased bacterial growth in the DWDS [9]. In this study, retention time had a significant impact on the bacterial community in the water, impacting both TCC and %HNA bacteria (Figure 6).

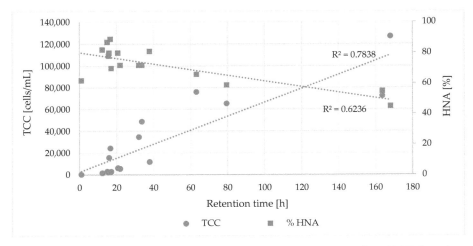

**Figure 6.** Correlation between retention time and TCC (green dots), as well as %HNA bacteria (blue squares).

An increased contact time between biofilm and water increased the number of bacteria (correlation coefficient $R = 0.89$; $p < 0.0001$). This could be due to an increased time during which the bacteria can detach from the biofilm and accumulate in the water phase, or an increased retention time could also provide bacteria in the water the chance to multiply [35]. Higher retention times were also correlated with a decrease in the %HNA bacteria, from about 80% for water with a retention time below 20 h to 50% when retention time exceeded 160 h (correlation coefficient $R = -0.79$; $p = 0.0003$). Differences in community structures for water samples with disparities in retention time and distance from the DWTP have been observed in previous studies [28]. One recent study showed a predominance of LNA bacteria in branch ends of DWDSs, which indicates similar trends compared to the results of this study [36]. Interestingly, this strong negative correlation between retention time and %HNA bacteria was not observed when %HNA were examined with respect to contact area. Studies show that LNA, as well as HNA, bacteria are able to grow in oligotrophic water [37]. Drinking water is an oligotrophic environment with total organic carbon (TOC) concentrations below 1 mg/L [36], and the DWDS in the current study was no exception: TOCeq measured in the permeate from the UF membrane showed stable and relatively low TOCeq values of about 1.83 ± 0.31 mg/L from April 2018 until April 2019 (Supplementary Materials Figure S8). Those values are comparable with the TOCeq values for the outgoing drinking water. This could explain why, in this DWDS, the %HNA bacteria decreased with retention time. While the specific growth rate of LNA bacteria is lower compared to the specific growth rate of HNA bacteria, suggesting that regrowth of HNA bacteria in the DWDS should be favored [37], LNA bacteria could predominate with increased retention time if the mechanism by which bacteria entering the water is not due to growth. This supports the conclusions made by Chan et al. that it is bacteria entering the water from the biofilm that are the predominant source of bacteria in this distributed water [2]. This is supported by the observation that contact area is only correlated with increasing TCC, and not with any change in %HNA, suggesting that the community of bacteria leaving the biofilm and entering the water has a consistent composition which is not altered by location of the biofilm in the DWDS. Perhaps the water quality in this DWDS selects which bacteria leave the biofilm at different locations, since presumably there are diverse biofilm populations within the DWDS, due to the presence of different pipe materials [38].

## 4. Conclusions

Using FCM, it was possible to capture the annual dynamics of the biofilm in this DWDS, quantifying changes in TCC, percentage of intact cells, and changes in the community composition, reflected in the amounts of LNA and HNA bacteria. The seasonal changes indicated an annual reproducibility and allowed the determination of baselines for different points in the DWDS.

The major findings of this study are:

1.  The bacterial community in the DWDS experienced clear seasonal changes with similar patterns for different areas of the DWDS. An increase in water temperature led to a significant increase in TCC during the summer period (range: 1.51–5.24-fold increase) at some locations.
2.  Hydraulic and specific pipe conditions influence the bacterial community in the water. FCM results indicated that an increase in retention time led to a decrease in the %HNA bacteria in the drinking water (correlation coefficient $R = -0.79$).
3.  Longer retention times and increased contact between the water and pipe biofilm led to an increase in TCC. Significant differences in the TCC could be seen in different areas of the DWDS depending on distance from the DWTP and retention time.

## 5. Limitations and Future Perspectives

In this, and other studies, the combination of various water treatments processes in the DWTP, and different conditions even within an individual DWDS, complicates the formulation of general statements regarding the microbiology in the water [1]. This necessitates the examination of each DWDS individually and the determination of baselines in a seasonal context. As changes of the bacteria in the drinking water are influenced by exchange between water and biofilm in the pipes [39] and regrowth due to changes in residual disinfectant concentration or substrate availability [40], determining how each factor contributes at different locations and seasons is difficult. One influence can be favored by different attributes of the heterogeneous features of the DWDS such as pipe dimensions, pipe material, and water path, as well as retention time which varies depending on local water demand [28]. In this context, the current study is limited, only indicating the complexity of the bacterial content in the water, and indirectly the impact of biofilm from the DWDS. Further investigations are needed to advance our understanding of, for example, the impact of nutrients on the biofilm, in order to generate new and detailed insights into the role played by microbiology in environmental engineering.

The main purpose of this study was to compile a comprehensive description of temporal and spatial bacterial dynamics in the drinking water, and indirectly, the biofilm. This resulted in the preliminary definitions of baselines for microbiological water quality in different zones in this DWDS, providing a new and essential strategy for enhanced drinking water management in this system. Future monitoring incorporating these baselines can be used to detect abnormalities and sudden changes in the bacterial content of the water that may occur due to malfunctions, water leakages, or pipe maintenance, as well as ensuring that values return to the baselines to indicate recovery of acceptable water quality in the DWDS. The possibility to rapidly identify changes in the number of bacteria in the water, and any altered pattern of regrowth, becomes even more important with regard to a possible removal of chloramine in this DWDS in the future. The use of FCM to monitor the status of the distributed water can contribute to managing the uncertainty of how the removal of the disinfectant will affect bacterial regrowth in the pipes, as results can be rapidly available and compared to robust baselines for surveillance. However, the lack of disinfectant residual makes proper engineering practices in operation, maintenance, and construction of distribution networks vital in order to protect the drinking water quality [41].

**Supplementary Materials:** The following are available online at http://www.mdpi.com/2073-4441/11/10/2137/s1: Figure S1: Old and new treatment process at Kvarnagården DWTP in Varberg, Sweden; Figure S2: Water sampling points in the DWDS in Varberg, Sweden; Figure S3: Schematic description of TCC in the DWDS, and increase of TCC and water temperature from spring to summer; Figure S4: Intact cell count in connection with residues of chloramine in the DWDS in mg/L; Figure S5: Changes in TCC, ICC, and water temperature at sampling point Hunst; Figure S6: Changes in TCC, ICC, and water temperature at sampling point TrPS5; Figure S7: Increase in TCC at sampling point TrPS5 in late June 2019; Figure S8: TOCeq measured in the permeate of the UF membrane from April 2018 to April 2019; Table S1:Flow cytometry results and environmental parameters for different sampling points.

**Author Contributions:** Conceptualization, All; methodology, All; formal analysis, C.S. and P.R.; investigation, C.S.; writing—original draft preparation, C.S., P.R., M.D.B. and A.K.; writing—review and editing, All; visualization, C.S., K.P. and S.C.; supervision, A.K.; project administration, A.K.; funding acquisition, A.K., P.R. and C.J.P.

**Funding:** This research received external financial support of the Biofilm study-project (Project no. 17–104) funded by The Swedish Water and Wastewater Association (Swedish Water Development, SVU).

**Acknowledgments:** The authors acknowledge the financial support from Vatten & Miljö I Väst AB, which funded this long-term study, including monitoring program and analyses.

**Conflicts of Interest:** The authors declare no conflicts of interest.

## References

1. Chan, S. Processes Governing the Drinking Water Microbiome. Ph.D. Thesis, Faculty of Engineering (LTH), Lund, Sweden, 2018.

2. Chan, S.; Pullerits, K.; Keucken, A.; Persson, K.M.; Paul, C.J.; Rådström, P. Bacterial release from pipe biofilm in a full-scale drinking water distribution system. *NPJ Biofilms Microbiomes* **2019**, *5*, 9. [CrossRef] [PubMed]

3. Liu, G.; Zhang, Y.; van der Mark, E.; Magic-Knezev, A.; Pinto, A.; van den Bogert, B.; Liu, W.; van der Meer, W.; Medema, G. Assessing the origin of bacteria in tap water and distribution system in an unchlorinated drinking water system by SourceTracker using microbial community fingerprints. *Water Res.* **2018**, *138*, 86–96. [CrossRef] [PubMed]

4. Waak, M.B.; Hozalski, R.M.; Hallé, C.; Lapara, T.M. Comparison of the microbiomes of two drinking water distribution systems-With and without residual chloramine disinfection. *Microbiome* **2019**, *7*. [CrossRef] [PubMed]

5. Feron, D.; Neumann, E. "Biocorrosion 2012"—From advanced technics towards scientific perspectives. *Bioelectrochemistry* **2014**, *97*, 1. [CrossRef]

6. Piriou, P.; Malleret, L.; Bruchet, A.; Kiéné, L. Trichloroanisole kinetics and musty tastes in drinking water distribution systems. *Water Supply* **2001**, *1*, 11–18. [CrossRef]

7. Van der Kooij, D.; van der Wielen, P.W.J.J. *Microbial Growth in Drinking-Water Supplies: Problems, Causes, Control and Research Needs*, 1st ed.; IWA Publishing: London, UK, 2013; pp. 1–32.

8. Baskerville, A.; Broster, M.; Fitzgeorge, R.B.; Hambleton, P.; Dennis, P.J. Experimental Transmission of Legionnaires' Disease by Exposure to Aerosols of Legionella Pneumophila. *Lancet* **1981**, *318*, 1389–1390. [CrossRef]

9. Bartram, J.; Cotruvo, J.A.; Exner, M.; Fricker, C.; Glasmacher, A. *Heterotrophic Plate Counts and Drinking-Water Safety-The Significance of HPCs for Water Quality and Human Health*, 1st ed.; IWA Publishing: London, UK, 2003; pp. 80–118.

10. Pinto, A.J.; Xi, C.; Raskin, L. Bacterial community structure in the drinking water microbiome is governed by filtration processes. *Environ. Sci. Technol.* **2012**, *46*, 8851–8859. [CrossRef]

11. Hijnen, W.A.M.; Beerendonk, E.F.; Medema, G.J. Inactivation credit of UV radiation for viruses, bacteria and protozoan (oo)cysts in water: A review. *Water Res.* **2006**, *40*, 3–22. [CrossRef]

12. Douterelo, I.; Boxall, J.B.; Deines, P.; Sekar, R.; Fish, K.E.; Biggs, C.A. Methodological approaches for studying the microbial ecology of drinking water distribution systems. *Water Res.* **2014**, *65*, 134–156. [CrossRef]

13. Hammes, F.; Berney, M.; Wang, Y.; Vital, M.; Köster, O.; Egli, T. Flow-cytometric total bacterial cell counts as a descriptive microbiological parameter for drinking water treatment processes. *Water Res.* **2008**, *42*, 269–277. [CrossRef]

14. Lautenschlager, K.; Boon, N.; Wang, Y.; Egli, T.; Hammes, F. Overnight stagnation of drinking water in household taps induces microbial growth and changes in community composition. *Water Res.* **2010**, *44*, 4868–4877. [CrossRef] [PubMed]

15. Payment, P.; Trudel, M.; Plante, R. Elimination of viruses and indicator bacteria at each step of treatment during preparation of drinking water at seven water treatment plants. *Appl. Environ. Microbiol.* **1985**, *49*, 1418–1428. [PubMed]

16. Van Nevel, S.; Buysschaert, B.; De Gusseme, B.; Boon, N. Flow cytometric examination of bacterial growth in a local drinking water network. *Water Environ. J.* **2016**, *30*, 167–176. [CrossRef]
17. Lautenschlager, K.; Hwang, C.; Liu, W.T.; Boon, N.; Köster, O.; Vrouwenvelder, H.; Egli, T.; Hammes, F. A microbiology-based multi-parametric approach towards assessing biological stability in drinking water distribution networks. *Water Res.* **2013**, *47*, 3015–3025. [CrossRef]
18. Van Nevel, S.; Koetzsch, S.; Proctor, C.R.; Besmer, M.D.; Prest, E.I.; Vrouwenvelder, J.S.; Knezev, A.; Boon, N.; Hammes, F. Flow cytometric bacterial cell counts challenge conventional heterotrophic plate counts for routine microbiological drinking water monitoring. *Water Res.* **2017**, *113*, 191–206. [CrossRef]
19. Gatza, E.; Hammes, F.A.; Prest, E.I. White Paper: Assessing Water Quality with the BD Accuri TM C 6 Flow Cytometer. 2013. Available online: https://www.semanticscholar.org/paper/White-Paper-Assessing-Water-Quality-with-the-BD-TM-Gatza-Hammes/8e96101b49bfcb09e28425cf40c0cce2dec1afca (accessed on 18 July 2019).
20. Besmer, M.D.; Sigrist, J.A.; Props, R.; Buysschaert, B.; Mao, G.; Boon, N.; Hammes, F. Laboratory-scale simulation and real-time tracking of a microbial contamination event and subsequent shock-chlorination in drinking water. *Front. Microbiol.* **2017**, *8*, 1900. [CrossRef]
21. Keucken, A.; Heinicke, G.; Persson, K.M.; Köhler, S.J. Combined coagulation and ultrafiltration process to counteract increasing NOM in brown surface water. *Water* **2017**, *9*, 697. [CrossRef]
22. Prest, E.I.; Hammes, F.; Kötzsch, S.; van Loosdrecht, M.C.M.; Vrouwenvelder, J.S. Monitoring microbiological changes in drinking water systems using a fast and reproducible flow cytometric method. *Water Res.* **2013**, *47*, 7131–7142. [CrossRef]
23. Chan, S.; Pullerits, K.; Riechelmann, J.; Persson, K.M.; Rådström, P.; Paul, C.J. Monitoring biofilm function in new and matured full-scale slow sand filters using flow cytometric histogram image comparison (CHIC). *Water Res.* **2018**, *138*, 27–36. [CrossRef]
24. Koch, C.; Fetzer, I.; Harms, H.; Müller, S. CHIC-an automated approach for the detection of dynamic variations in complex microbial communities. *Cytom. Part A* **2013**, *83*, 561–567. [CrossRef]
25. Ellis, B.; Haaland, P.; Hahne, F.; Meur, N.L.; Gopalakrishnan, N.; Spidlen, J.; Jiang, M. FlowCore: Basic Structures for Flow Cytometry Data, Bioconductor R. Package Version 1.40.0. 2016. Available online: https://rdrr.io/bioc/flowCore/ (accessed on 7 August 2019).
26. Oksanen, J.; Blanchet, F.G.; Kindt, R.; Legendre, P.; Minchin, P.; O'Hara, R.B.; Simpson, G.; Solymos, P.; Stevens, M.H.H.; Wagner, H. Vegan: Community Ecology Package. R Package Version 2.2-1. 2015. Available online: http://CRAN.R-project.org/package=vegan (accessed on 14 August 2019).
27. Keucken, A. Climate Change Adaption of Waterworks for Browning Surface Waters: Nano- and Ultrafiltration Membrane Applications for Drinking Water Treatment. Ph.D. Thesis, Faculty of Engineering (LTH), Lund, Sweden, 2017.
28. Pinto, A.J.; Schroeder, J.; Lunn, M.; Sloan, W.; Raskin, L. Spatial-temporal survey and occupancy-abundance modeling to predict bacterial community dynamics in the drinking water microbiome. *mBio* **2014**, *5*. [CrossRef] [PubMed]
29. Liu, S.; Gunawan, C.; Barraud, N.; Rice, S.A.; Harry, E.J.; Amal, R. Understanding, monitoring, and controlling biofilm growth in drinking water distribution systems. *Environ. Sci. Technol.* **2016**, *50*, 8954–8976. [CrossRef] [PubMed]
30. Prest, E.I.; Weissbrodt, D.G.; Hammes, F.; Van Loosdrecht, M.C.M.; Vrouwenvelder, J.S. Long-term bacterial dynamics in a full-scale drinking water distribution system. *PLoS ONE* **2016**, *11*, e0164445. [CrossRef] [PubMed]
31. Douterelo, I.; Sharpe, R.L.; Boxall, J.B. Influence of hydraulic regimes on bacterial community structure and composition in an experimental drinking water distribution system. *Water Res.* **2013**, *47*, 503–516. [CrossRef] [PubMed]
32. Lipphaus, P.; Hammes, F.; Kötzsch, S.; Green, J.; Gillespie, S.; Nocker, A. Microbiological tap water profile of a medium-sized building and effect of water stagnation. *Environ. Technol.* **2014**, *35*, 620–628. [CrossRef] [PubMed]
33. Chiao, T.H.; Clancy, T.M.; Pinto, A.; Xi, C.; Raskin, L. Differential resistance of drinking water bacterial populations to monochloramine disinfection. *Environ. Sci. Technol.* **2014**, *48*, 4038–4047. [CrossRef]
34. Boe-Hansen, R.; Albrechtsen, H.J.; Arvin, E.; Jørgensen, C. Bulk water phase and biofilm growth in drinking water at low nutrient conditions. *Water Res.* **2002**, *36*, 4477–4486. [CrossRef]

35. Universität Göttingen. Available online: https://lp.uni-goettingen.de/get/text/4908 (accessed on 17 June 2019).

36. Liu, J.; Zhao, Z.; Chen, C.; Cao, P.; Wang, Y. In-situ features of LNA and HNA bacteria in branch ends of drinking water distribution systems. *J. Water Supply Res. Technol. AQUA* **2017**, *66*, 300–307. [CrossRef]

37. Wang, Y.; Hammes, F.; Boon, N.; Chami, M.; Egli, T. Isolation and characterization of low nucleic acid (LNA)-content bacteria. *ISME J.* **2009**, *3*, 889–902. [CrossRef]

38. Ren, H.; Wang, W.; Liu, Y.; Liu, S.; Lou, L.; Cheng, D.; He, X.; Zhou, X.; Qiu, S.; Fu, L.; et al. Pyrosequencing analysis of bacterial communities in biofilms from different pipe materials in a city drinking water distribution system of East China. *Appl. Microbiol. Biotechnol.* **2015**, *99*, 10713–10724. [CrossRef]

39. Henne, K.; Kahlisch, L.; Brettar, I.; Höfle, M.G. Analysis of structure and composition of bacterial core communities in mature drinking water biofilms and bulk water of a citywide network in Germany. *Appl. Environ. Microbiol.* **2012**, *78*, 3530–3538. [CrossRef] [PubMed]

40. Srinivasan, S.; Harrington, G.W.; Xagoraraki, I.; Goel, R. Factors affecting bulk to total bacteria ratio in drinking water distribution systems. *Water Res.* **2008**, *42*, 3393–3404. [CrossRef] [PubMed]

41. Medema, G.J.; Smeets, P.W.; Blokker, E.J.; van Lieverloo, J.H. Safe distribution without a disinfectant residual. In *Microbial Growth in Drinking-Water Supplies: Problems, Causes, Control and Research Needs*, 1st ed.; van der Kooij, D., van der Wielen, P.W.J.J., Eds.; IWA Publishing: London, UK, 2013; pp. 95–125.

Article

# The Impact of the Quality of Tap Water and the Properties of Installation Materials on the Formation of Biofilms

Dorota Papciak [1], Barbara Tchórzewska-Cieślak [2], Andżelika Domoń [1,*], Anna Wojtuś [1], Jakub Żywiec [2] and Janusz Konkol [3]

[1]   Department of Water Purification and Protection, Faculty of Civil, Environmental Engineering and Architecture, Rzeszow University of Technology, Al. Powstancow Warszawy 6, 35-959 Rzeszow, Poland; dpapciak@prz.edu.pl (D.P.); annawojtusss@gmail.com (A.W.)
[2]   Department of Water Supply and Sewerage Systems, Faculty of Civil, Environmental Engineering and Architecture, Rzeszow University of Technology, Al. Powstancow Warszawy 6, 35-959 Rzeszow, Poland; cbarbara@prz.edu.pl (B.T.-C.); j.zywiec@prz.edu.pl (J.Ż.)
[3]   Department of Materials Engineering and Technology of Building, Faculty of Civil, Environmental Engineering and Architecture, Rzeszow University of Technology, Al. Powstancow Warszawy 6, 35-959 Rzeszow, Poland; janusz.konkol@prz.edu.pl
*   Correspondence: adomon@prz.edu.pl; Tel.: +48-17-865-1949

Received: 8 May 2019; Accepted: 9 September 2019; Published: 12 September 2019

**Abstract:** The article presents changes in the quality of tap water depending on time spent in installation and its impact on the creation of biofilms on various materials (polyethylene (PE), polyvinyl chloride (PVC), chrome-nickel steel and galvanized steel). For the first time, quantitative analyses of biofilm were performed using methods such as: Adenosine 5'-triphosphate (ATP) measurement, flow cytometry, heterotrophic plate count and using fractographical parameters. In the water, after leaving the experimental installation, the increase of turbidity, content of organic compounds, nitrites and nitrates was found, as well as the decrease in the content of chlorine compounds, dissolved oxygen and phosphorus compounds. There was an increase in the number of mesophilic and psychrophilic bacteria. In addition, the presence of *Escherichia coli* was also found. The analysis of the quantitative determination of microorganisms in a biofilm indicates that galvanized steel is the most susceptible material for the adhesion of microorganisms. These results were also confirmed by the analysis of the biofilm morphology. The roughness profile, the thickness of the biofilm layer can be estimated at about 300 μm on galvanized steel.

**Keywords:** biofilm; tap water; water supply system

## 1. Introduction

Water supply systems in residential buildings, public utilities or industrial plants are made of various materials [1]. The type of material and the quality of tap water are the most important factors affecting the risk of losing water safety reaching the consumer [2]. Therefore, the assessment of risks related to internal water supply installations, including hazards related to products and materials in contact with drinking water is extremely important [3]. A particular health risk is caused by the emergence of biofilm and increased colonization of organisms on the surface of water pipes. The intensity of biofilm growth in a water distribution system depends on numerous factors such as content of biogenic compounds in water injected into a network, amount of disinfectant, temperature, hydrodynamic conditions and the type of material from which the water conduits are made [4–8]. The chemical composition of the water supply system material, as well as its properties, that is, porousness and susceptibility to corrosion, are regarded as one of the main causes of increased colonization of microorganisms. The multitude of factors influencing the formation and development of biofilm makes prevention of the phenomenon very complex [9–11].

Plastics and polyester resins in tap water distribution networks are replacing cast-iron and galvanized steel, however; pipes and elements made of copper or chromium-nickel steel are used as well. These materials are characterized by high resistance to corrosion and low surface porousness. Due to their ever-reducing price and valuable properties, they can become genuine competition for traditional materials used in water supply systems (such as polyethylene (PE), and polyvinyl chloride (PVC)). Currently, cast-iron and concrete are used decidedly less often to build water-pipe networks than just a few years ago, however; numerous sections of networks made of these materials are still in exploitation [12].

Synthetic materials characterized by low porousness were expected to eliminate corrosion and decrease the risk of secondary water pollution. The research conducted thus far shows that biofilm forms on all water supply system materials, but each of them creates different conditions for duplication and adhesion of microorganisms in the form of biofilm [13].

According to the prevailing literature on the issue, plastics can support the formation of biofilm, however, the growth of microorganisms on the surface of plastic pipes is usually the same or lower than in the case of iron, steel, or concrete [4,7].

Contact between water and the internal surface of the water conduits may lead to, i.e., corrosion or ageing of materials, eluting chemical substances, and to the formation of biofilm on their surfaces. Metals from brass elements, e.g., joints, can also be transferred into the water. This applies particularly to lead, which presents a serious risk to pregnant women and children below six years of age [1].

Phosphorus and carbon, which are nutrients for microorganisms, can transfer to water when materials come into contact with them in the form of microbiologically available phosphorus (MAP) [5] or available organic carbon (AOC) [14], speeding up biofilm formation.

Despite the possible influence of material on biofilm formation [15–17], in research conducted on actual conduit sections after a year of contact with tap water, this dependency has not been noted [1]. It was determined, however, that the biofilm formation process was influenced by water temperature and flow conditions [1].

Some researchers state that type of material influences biofilm formation, but only at its initial stage. According to the research, the small difference in the number of microorganisms populating internal surfaces of PE and copper pipes after 21 and 200 days shows the lack of dependency between material type and intensity of biofilm growth after a longer time period in contact with the medium [5].

None of the so far examined materials allows the complete elimination of biofilm formation in tap water distribution networks. It is, therefore, noteworthy that, due to their structure, both plastics and corroding materials create different opportunities for the formation of biofilm [5,13,16,18].

The research so far has focused on specific materials and selected different conditions of biofilm formation on material surfaces [1,4,5,18,19], experiment time duration [1,4,5,19,20], biofilm detachment method for analysis [4,5,18–20] and biofilm quantification methods [1,4,5,18,19]. Therefore, this article presents an evaluation method of the susceptibility of water supply system materials to biofilm formation in conditions close to those of actual water supply systems.

An additional possibility of analyzing biofilm morphology is provided by the use of fractal analysis together with the appropriate tools of fractal geometry [21] used to quantify the roughness of any structure. The biofilm research in such a wide range has not been conducted so far.

The aim of the study was to assess changes in water quality and biological stability depending on the time spent in the distribution system and to determine the susceptibility of materials to the adhesion of microorganisms.

## 2. Materials and Methods

*2.1. Subject of Study*

The experimental water supply system consisted of three main parts: A tap water connection, circulation in the experimental system, and a discharge of water from the system (Figure 1). Two ball valves and a water meter measuring the fresh tap water flow were installed on the connection conduit. The experimental water supply system was a closed circuit made of PVC DN (Diameter Nominal) 32 pipes, in which circulation was ensured by an installed pump. The circuit was equipped with a separate water meter, drain and vent lines, flow decrease installation, and disinfectant injection point. The samples of examined materials (PVC, PE, chromium-nickel steel, galvanized steel) attached with a stainless metal rod to PVC plugs along with an O-ring rubber seal, were mounted on specially modified tees separated with ball valves (Figure 2). The discharge of water from the system was obtained via a combination of drain and vent lines. Water flow speed in the experimental system equaled 0.3 m/h. A collection of samples for physical, chemical, and microbiological tests was via a tap with a cable, installed on the conduit discharging water from the system to the sewage system.

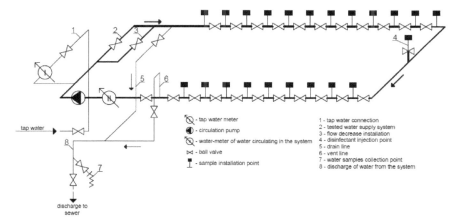

Figure 1. The scheme of the experimental installation.

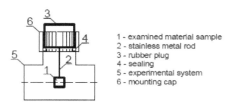

Figure 2. The scheme of the tested sample.

In the experimental installation, all analyzed installation materials were installed. Five coupons were prepared for each material, placed in the following order: PE, PVC, chromium-nickel steel, galvanized steel. Fresh tap water was introduced into the installation once a day, and then the water circulated in the closed circuit in the installation for 24 hours. Before starting the tests, the installation together with the coupons was disinfected with 15% sodium hypochlorite.

The building with the installation was located 5 km from the water treatment plant.

## 2.2. Water in the Experimental Installation

The experimental system was supplied with surface water which had been treated using the technology presented in Figure 3 and met the quality requirements of water intended for human consumption. Tests on quality changes in the water leaving and supplying the experimental system were conducted once a week for 6 months. Selected physical, chemical, and bacteriological parameters of water supplying and leaving the experimental system were estimated using the methods presented in Table 1. The tests were carried out in accordance with the applicable test procedures and the manufacturer's instructions were followed.

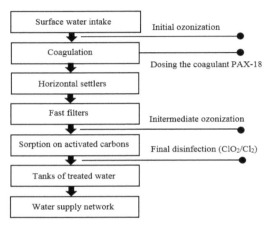

**Figure 3.** Water treatment technology.

**Table 1.** Analytical methods and standards used in experiment.

| Parameter | Analytical Method/Standard |
|---|---|
| pH | Multifunction meter CX-505 (Elmetron, Poland) |
| Temperature | Multifunction meter CX-505 (Elmetron, Poland) |
| Conductivity | Multifunction meter CX-505 (Elmetron, Poland) |
| Turbidity | 2100P ISO turbidimeter (Hach, Germany) |
| Oxidizability | The permanganate method with $KMnO_4$ (according PN EN ISO 8467:2001) |
| Total organic carbon (TOC) | TOC analyzer Sievers 5310 C (SUEZ, Boulder, CO, USA) |
| UV absorbance | Spectrophotometric method using Hach–Lange DR 500 spectrophotometer (Germany) |
| Dissolved oxygen | Electrochemical method using a Hach–Lange oxygen probe (Germany) |
| Ammonium nitrogen | Spectrophotometric method 8155 (sachet tests—amonia salicylate (1) and cyanurate (2)) using Hach-Lange DR 500 spectrophotometer (Germany) |
| Nitrite nitrogen | Colorimetric method by Nitrite Test Merck 1.14408 (Germany) |
| Nitrate nitrogen | Spectrophotometric method 8039 (sachet tests—NitraVer5) using Hach–Lange DR 500 spectrophotometer (Germany) |
| Phosphates | Spectrophotometric method 8048 (sachet tests—PhosVer3) using Hach–Lange DR 500 spectrophotometer (Germany) |
| Total and free chlorine | Spectrophotometric method 8167 and 8021 (sachet tests—DPD reagent) using Hach-Lange DR 500 spectrophotometer (Germany) |
| The total number of psychrophilic bacteria (at 22 °C) and mesophilic bacteria (at 37 °C) | Heterotrophic plate count (HPC) method using R2A Agar (CM0906) manufactured by Oxoid Thermo Scientific (UK) (incubation 7 days) and an A Agar (P-0231) manufactured by BTL Sp. z o.o. Department of Enzymes and Peptones (Poland) (incubation 2 day-mesophilic bacteria and 3 day psychrophilic bacteria) |
| *Escherichia coli* | Membrane filtration procedure using Endo agar WG ISO 9308-1 (BTL, Poland) |

Biodegradable dissolved organic carbon (BDOC) content was calculated on the basis of data published by Wolska, who determined that in the case of surface waters, the dominant fraction of dissolved organic carbon (DOC) was the non-biodegradable fraction and equaled 90% of total organic carbon (TOC), while BDOC content equaled ca. 10.6% of DOC [11].

### 2.3. The Susceptibility of Materials to the Formation of Biofilms in the Experimental Installation

The assessment of susceptibility of materials to microorganisms adhesion was conducted on the basis of results of microbiological tests by ATP (Adenosine 5'-triphosphate) amount measurement (using LuminUltra Photonmaster Luminometer), flow cytometry method (using a Cy Flow Cube 8 cytometer manufactured by Sysmex Partec), and HPC methods with A and R2A agars. Biofilm detachment from the material surface was obtained using a 60-second sonication. Coupons of examined materials were placed in the water supply system for 6 months (December 2017–June 2018).

In order to determine the susceptibility of materials to the formation of biofilms, fractographical parameters were used. The measuring apparatus included Taylor Hobson laser profilometer Talysurf CLI 1000 for fast non-contact measurement of 3D surface topography together with TalyMap and FRAKTAL_Dimmension2D software. The fractal study was conducted on surfaces of selected materials (galvanized steel and PVC) using a box counting method. The method consists of enclosing each section of a profile by a box of width $\varepsilon$ and calculating the area $N(\varepsilon)$ of all of the boxes enclosing the whole profile. This procedure is iterated with boxes of different widths to build a graph; $\ln(N(\varepsilon))/\ln(\varepsilon)$.

In addition, the total height of the roughness profile (Pt) was determined for each profile. The profiled lines with a length of 5 mm separated from the surface of the tested materials were determined with a discretization step of 0.5 µm. The number of profile lines designated on the surface of a given material was 12, which can be considered as adequate [22].

## 3. Results

### 3.1. Water Quality Assessment

Water leaving the experimental system had different parameters in comparison to water supplying it (Table 2). As a result of contact with PVC material, an increase of turbidity and organic compounds content (TOC, oxidizability, UV absorbance) as well as a small increase of nitrites were noted, while ammonium nitrogen content remained unchanged. Simultaneously, content of chlorine (total and free) compounds, dissolved oxygen, and phosphorus compounds decreased. The bacteriological quality of the water also changed; the number of mesophilic and psychrophilic bacteria increased and the presence of *Escherichia coli* (0–48 CFU/100mL) was noted. The presence of bacteria could have been caused by the following: Contamination during assembly, insufficient rinsing, and disinfection of the installation. The high temperature prevailing in the installation (21.53 °C), and the washing out of nutrients from PVC additionally favored the multiplication of microorganisms. Changes to the parameters of water leaving the system pointed to biological processes occurring inside the conduits.

The data presented in Table 3 proves that, in water injected into the experimental system, the admissible content of nutritional substrates N, P, and C was exceeded [23], which could have had a significant influence on the formation of biofilm on the surface of examined materials.

**Table 2.** Inlet and outlet water quality characteristics (N = 26).

| Parameter | Unit | Inlet | | | | Outlet | | | |
|---|---|---|---|---|---|---|---|---|---|
| | | Min | Max | Mean | σ | Min | Max | Mean | σ |
| pH | - | 7.01 | 7.69 | 7.54 | 0.17 | 7.17 | 7.74 | 7.60 | 0.14 |
| Temperature | °C | 14.47 | 20.3 | 17.75 | 1.99 | 10.09 | 23.9 | 21.53 | 3.54 |
| Conductivity | μs/cm | 383 | 506 | 430 | 35.31 | 475 | 662 | 543 | 53.07 |
| Turbidity | NTU | 0.16 | 1.33 | 0.40 | 0.35 | 0.58 | 4.5 | 1.41 | 1.08 |
| Oxidizability | mg $O_2$/L | 0.50 | 2.10 | 1.41 | 0.47 | 0.80 | 2.60 | 1.73 | 0.56 |
| TOC | mg C/L | 0.98 | 2.05 | 1.52 | 0.26 | 1.99 | 5.00 | 2.44 | 0.86 |
| UV absorbance | $UV_{254\,nm}$ | 1.48 | 2.76 | 2.15 | 0.35 | 2.42 | 3.70 | 2.89 | 0.60 |
| Dissolved oxygen | mg $O_2$/L | 12.56 | 16.30 | 14.32 | 1.13 | 5.83 | 10.25 | 9.25 | 1.23 |
| Ammonium nitrogen | mg $N-NH_4^+$/L | 0.00 | 0.070 | 0.018 | 0.028 | 0.00 | 0.11 | 0.018 | 0.031 |
| Nitrite nitrogen | mg $N-NO_2^-$/L | 0.00 | 0.037 | 0.003 | 0.010 | 0.001 | 0.037 | 0.0071 | 0.04 |
| Nitrate nitrogen | mg $N-NO_3^-$/L | 0.09 | 0.90 | 0.49 | 0.29 | 0.20 | 1.50 | 0.52 | 0.376 |
| Phosphates | mg $PO_4^{3-}$/L | 0.02 | 0.19 | 0.053 | 0.047 | 0.00 | 0.15 | 0.038 | 0.037 |
| Total chlorine | mg $Cl_2$/L | 0.01 | 0.21 | 0.102 | 0.07 | 0.01 | 0.07 | 0.027 | 0.017 |
| Free chlorine | mg $Cl_2$/L | 0.01 | 0.08 | 0.033 | 0.02 | 0.00 | 0.04 | 0.012 | 0.011 |
| Mesophilic bacteria (R2A) | CFU/mL | 1 | 100 | 30 | 34 | 300 | 5200 | 2393 | 1561 |
| Psychrophilic bacteria (R2A) | CFU/mL | 5 | 90 | 49 | 32 | 450 | 10600 | 4401 | 3721 |
| *Escherichia coli* | CFU/100mL | 0.00 | 0.00 | 0.00 | 0.00 | 3.00 | 200.00 | 48 | 57.70 |

**Table 3.** Limit values and average concentrations of nutritional substrates in the studied waters (N = 26).

| Stability Criterion | Water Treatment Plant | Inlet (24 h) | Outlet (24 h) |
|---|---|---|---|
| | Mean | | |
| $\Sigma N_{inorg} \leq 0.2$ mg N/L | 0.930 | 0.510 | 0.540 |
| $PO_4^{3-} \leq 0.03$ mg $PO_4^{3-}$/L | 0.027 | 0.053 | 0.038 |
| Dissolved organic carbon (DOC) mg C/L | 2.160 | 1.520 | 2.440 |
| Biodegradable dissolved organic carbon (BDOC) $\leq 0.25$ mg C/L | 0.140 | 0.220 | 0.300 |

*3.2. Analysis of the Surface of Installation Materials*

After six months of contact with tap water, the surface of the examined materials was covered in biofilm. The number of microorganisms populating the material surface varied depending on their type (chemical composition), surface porousness, and method of microbiological estimation (Table 4). Agar R2A stimulated growth of a significantly higher number of microorganisms in comparison to agar A (at incubation times of two and three days), and the obtained results of the two surfaces differed by two orders of magnitude. In the case of agar R2A, both nutrient composition and incubation time (seven days) were different. It should also be noted that the sonication process used during biofilm removal could contribute to damaging part of the microorganisms gathered on the surface of examined materials. A longer incubation period could ensure better conditions for growth of damaged and stressed microorganisms.

Regardless of the method of bacteriological estimation, galvanized steel and PE were the most susceptible to microorganism adhesion materials, followed by chromium-nickel steel and PVC.

The number of mesophilic and psychrophilic microorganisms was comparable for PVC and chromium-nickel steel. In the case of galvanized steel and PE, two times more psychrophilic bacteria than mesophilic bacteria were noted (Table 4). Psychrophilic bacteria are a group of bacteria living and reproducing at low temperatures (0–25 °C) and are mostly Gram-negative G (−). The majority of pathogenic bacteria, as well as ground and water bacteria, belong to mesophilic bacteria [24]. The population size of mesophilic microorganisms provides information on the presence of pathogenic and potentially pathogenic microorganisms, while the population size of psychrophilic bacteria points to organic matter content [25]. Measurements taken via HPC methods, luminometric measurement of ATP, and flow cytometry cannot be compared to one another with regards to the number of microorganisms due to differences between estimated forms (colonies, ATP, living/dead particles). The obtained data also cannot be compared to values published by other authors, due to differences in the methodology of the conducted experiments (flow and static conditions; different times of biofilm formation, different temperature and medium composition, conditions of removal and estimation of microorganism numbers in the biofilm).

**Table 4.** Total number of microorganisms in the biofilm detachment from the surface of various materials.

| Material | The Number of Microorganisms | | | |
|---|---|---|---|---|
| | Agar A (CFU/cm$^2$) | Agar R2A (CFU/cm$^2$) | ATP (RLU/cm$^2$) | Flow Cytometry (Number of Particles/cm$^2$) |
| Galvanised steel | M 35, P 170 | M 9900, P 18,950, | 17,390 | 7,951,795 |
| PE | M 60, P 75 | M 9750, P 18,400 | 8507 | 7,992,750 |
| Chromium-nickel steel | M 53, P 45 | M 5200, P 5800 | 5650 | 7,341,230 |
| PVC | M 80, P 75 | M 3300, P 3200 | 4523 | 7,019,205 |

M—Mesophilic bacteria; P—psychrophilic bacteria; ATP—adenosine 5'-triphosphate; PE—polyethylene; PVC—polyvinyl chloride.

At this stage of our research, we can merely order the materials due to their susceptibility to biofilm formation. Lethola et al. points to copper as the material most resistant to biofilm and states that this material influences biofilm formation solely at the initial stage of contact between water and the material [5]. Copper is often used for in-building systems, the ions of which are toxic to bacteria thus decreasing potential biofilm formation. On the other hand, it should be noted that the material characterizes with high porousness in comparison to plastics, due to which it creates good conditions for colonization of microorganisms [26–28]. In the research [13,18], copper was noted as the least susceptible to biofilm material in comparison to stainless steel and plastics. Our observations suggest that eluting organic matter and biogenic compounds (phosphorus) from the system occurred (Tables 2 and 3) and could stimulate biofilm formation. Yu et al. also points to copper as the most resistant material, followed by PE and galvanized steel [18], which agrees with the order determined by our research: Galvanized steel > PE > chromium-nickel steel > PVC.

The examined chromium-nickel steel characterizes with low surface porousness and high resistance to corrosion. These features could have had a significant influence on the obtained test results. This material characterized with a lower susceptibility to biofilm than steel and PE. However, the main drawback of chromium-nickel is its susceptibility to the presence of chlorine and strong oxidants in water, which translates into formation of pitting corrosion. Protection from this type of undesired situation and the high microbiological quality of water can be achieved through the addition of molybdenum or titan [29]. It is note-worthy that chromium-nickel is a relatively expensive material but, due to its increasing popularity and market competition, its price is systematically decreasing. Surprisingly, PVC turned out to be the most resistant to biofilm material and the cause of this should be examined in further research. Waller et al. conducted research in flow conditions using PVC and cast-iron coupons but the medium circulating in the system was a solution enriched with biogenic substances. Due to the method of biofilm removal from the surface of the examined materials, and more specifically, due to the type of physiological liquid used (phosphate-buffered saline liquid) [19], as well as different media coming into contact with the examined samples, there is no option of direct comparison of obtained test results. Nonetheless, in the case of HPC analysis of agar R2A for the highest number of bacteria, Waller obtained samples from iron (about 8,000,000 (CFU/cm$^2$)) after two days of contact between material and solution, while for PVC, the value of which equaled (80,000 (CFU/cm$^2$)), the samples were obtained after one day.

The results of the fractal analysis of the fractal dimension D values and the total height of the roughness profile together with the error of the standard mean value are summarized (Table 5,

Figures 4–11). The values of both parameters are given as a result of the analysis of the reference material surface and the biofilm material.

**Table 5.** Morphology of the surface of the reference material and material with biofilm.

| Material | Fractographical Parameters | | | |
|---|---|---|---|---|
| | Fractal Dimension D ± Standard Error (-) | | Total Height of the Roughness Profile $P_t$ ± Standard Error (μm) | |
| | For Material | For Material with Biofilm | For Material | For Material with Biofilm |
| Galvanised steel | 1.23 ± 0.016 | 1.18 ± 0.015 | 83.6 ± 6.2 | 393.9 ± 23.4 |
| PE | 1.40 ± 0.015 | 1.41 ± 0.011 | 39.4 ± 2.7 | 93.4 ± 2.9 |
| Chromium-nickel steel | 1.43 ± 0.006 | 1.35 ± 0.006 | 53.6 ± 3.0 | 114.7 ± 4.3 |
| PVC | 1.40 ± 0.029 | 1.39 ± 0.004 | 42.0 ± 1.4 | 30.0 ± 1.3 |

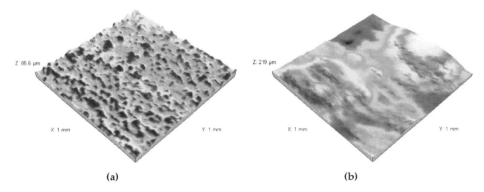

(a)                                    (b)

**Figure 4.** The surface of the reference material (galvanized steel) (**a**) and material covered with biofilm (**b**).

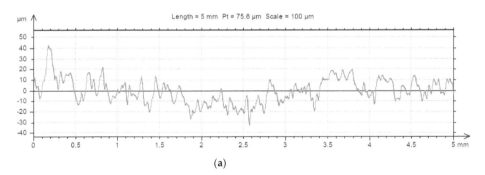

(a)

**Figure 5.** *Cont.*

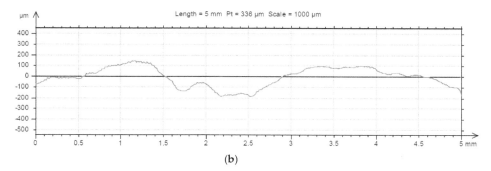

**Figure 5.** Representative profile lines for the reference material (galvanized steel) (**a**) and biofilm coated material (**b**).

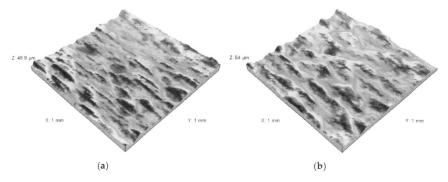

(**a**)          (**b**)

**Figure 6.** The surface of the reference material (PE) (**a**) and material covered with biofilm (**b**).

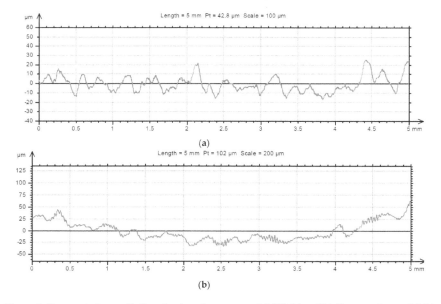

**Figure 7.** Representative profile lines for the reference material (PE) (**a**) and biofilm coated material (**b**).

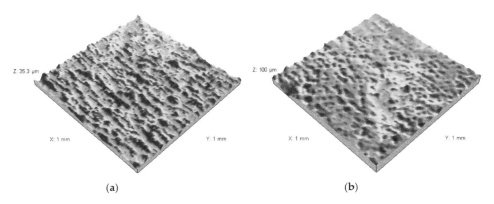

**Figure 8.** The surface of the reference material (chromium-nickel steel) (**a**) and material covered with biofilm (**b**).

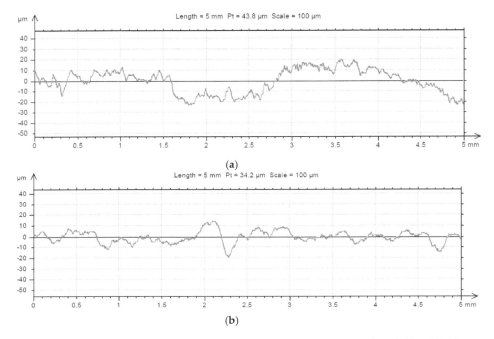

**Figure 9.** Representative profile lines for the reference material (chromium-nickel steel) (**a**) and biofilm coated material (**b**).

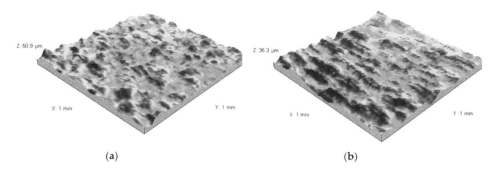

(a)                                                      (b)

**Figure 10.** The surface of the reference material (PVC) (**a**) and material covered with biofilm (**b**).

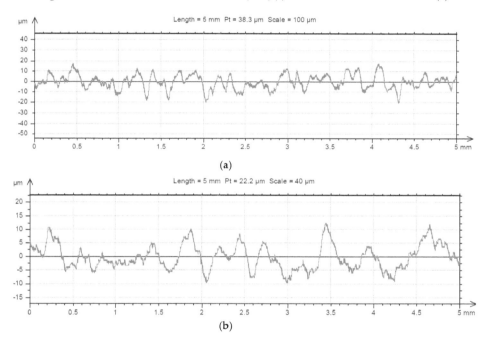

(a)

(b)

**Figure 11.** Representative profile lines for the reference material (PVC) (**a**) and biofilm coated material (**b**).

## 4. Discussion

In earlier research, Pietrucha et al. confirmed that tap water supplying the experimental system is chemically unstable water with a tendency to solve solids with the possibility of occurrence of slight corrosion [30].

Corrosion encourages biofilm development [31] and presents an additional risk of loss of physical and biological stability of tap water [10]. Striving to achieve biological stability of water directed to distribution networks is connected to the necessity of ensuring extremely low content of nutrients for microorganisms developing on the surfaces of the water-pipe network. This is a very difficult task, especially in the case of water treated in conventional systems (chemical oxidation, coagulation, filtration, disinfection). For the purpose of ensuring biological stability of water, effective elimination of organic substances and biogenic elements, nitrogen and phosphorus, is necessary. In order to maintain stability and maximally limit the risk of secondary biological water pollution, two out of three biogenes determining microorganism growth should be removed [10]. Zamorska in her research, confirmed the lack of biological stability of water injected into the water-pipe network [32].

In the case of water containing natural organic matter and inorganic nitrogen, phosphorus ions are crucial [2,17]. Too low content hinders microorganism development at a significantly higher degree than in the case of other biogenes [33]. Lehtola et al. suggest that due to the lowest required phosphorus content, it is this element that limits microorganism growth [34]. It should be noted that phosphorus and other nutrients may, in the first days of exploitation of water systems made of plastics, be eluting from these materials, causing quicker development of biofilm, which is what may have occurred in the described case [5,6,18].

Thresholds of parameters limiting redevelopment of microorganisms in distribution networks should be lower than 0.25 mg C/L BDOC, 0.2 mg $N_{norg}$/L and 0.03 mg $PO_4^{3-}$/L [11]. In the first few days after launching the system, concentration of $PO_4^{3-}$ ions increased systematically from 0.05 to 0.15 mg $PO_4^{3-}$/L. No further increase of the analyzed parameter was noted after 26 days (Figure 12). The phosphorus eluted into the water in lower quantities could be used by increased numbers of bacteria developed in the system. Similar research results were noted by [5]. Nonetheless, in conduits made of plastics, small amounts of phosphorus may be present for as long as 200 days [5]. A similar dependency was noted in the case of total organic carbon (TOC), the content of which, after passing through the system, increased from 2.03 mg C/L to 2.93 mg C/L.

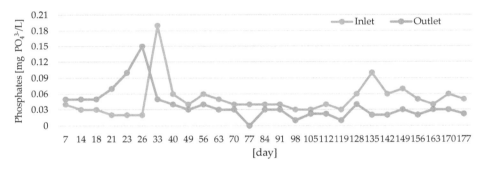

**Figure 12.** Change of phosphate content in the inflow and outflow water from the experimental installation.

Decrease of biofilm growth depends on water temperature, time duration of presence in the system, as well as type and concentration of disinfectant [32,35,36].

Due to the fact that the research was conducted in winter and spring, water temperature had a strong influence on microorganism development. A strong positive correlation between total number of psychrophilic bacteria and temperature was determined [37]. The temperature of water supplying the system changed in the range between 14.6 and 20.3 °C. Additionally, after passing through the system, the water temperature increased maximally up to 24 °C. According to the literary data, microorganism metabolic activity at 7 °C is lower by 50% than at 17 °C [38]. Total and elementary chlorine content in the water leaving the experimental system in comparison to water supplying the system decreased to 0.029 and 0.014 mg $Cl_2$/L, respectively. Chlorine is a disinfectant and a guarantee of the microbiological safety of water, which in this case, was clearly put at risk.

In Poland, the required concentration of elementary chlorine in water injected into the water-pipe network should equal 0.2–0.5 mg $Cl_2$/L, and in water at the ends of the network, it can be no lower than 0.05 g $Cl_2$/L. Gillespie et al. proved that systems distributing water with elementary chlorine concentrations below 0.5 mg $Cl_2$/L were associated with a larger number of bacteria cells in water [39]. Francisque et al. also shows that the number of heterotrophic bacteria was much higher in water samples with chlorine content < 0.3 mg $Cl_2$/L [40].

The proposed new method of assessing the adhesion of microorganisms based on the fractographic analysis of the material surface has proved to be a useful tool in the quantitative description of the biofilm structure.

The obtained values of the fractal dimension D (Table 5) indicate a greater surface roughness of the PVC material than galvanized steel material, which may result in a greater possibility of bacteria deposition in the depressions of this material. As shown by the results of microbiological tests and fractographical analyses, the dominant factor that indicates the possibility of biofilm formation is the type of material, not the roughness of its surface.

The reduction in the fractal dimension of the surface of galvanized steel with biofilm (D = 1.18) compared to the fractal dimension of the surface of this reference material (D = 1.23) indicates the deposition of biological material on this surface. The biofilm stratification is visible on an exemplary scanned fragment of material surface (Figure 4). In addition, the analysis of the shape of the profile lines and the definite change in total height of the roughness profiles indicate a significant occurrence of the biofilm on galvanized steel (Table 5, Figure 5). On the basis of total height of the roughness profile, the thickness of the biofilm layer can be estimated to be around 300 μm (the difference between the total height of the roughness profile of the biofilm material and the reference material).

A similar biofilm formation mechanism was found for chromium-nickel steel. The reduction in the fractal dimension of the surface of chromium-nickel steel with biofilm (D = 1.35) compared to the fractal dimension of the surface of this reference material (D = 1.43) indicates the deposition of biological material on this surface (Table 5, Figure 8). On the basis of total height of the roughness profile, the thickness of the biofilm layer can be estimated to be around 60 μm (Table 5, Figure 9).

In the case of PVC material, the fractal dimension does not show statistically significant (on the significance level of 0.05), changes in the roughness of the profile line. The changes are seen on exemplary fragments of the surface of the material (Figure 10) and in the total height of the roughness profile (Table 5, Figure 11). The reduction of the total height of the roughness profile indicates the formation of a biofilm in the material cavities. As can be seen in Figure 11, the structure is clearly orientated in the direction of the water flow.

No statistically significant difference was also found for PE material. In this case, an increase in total height of the roughness profile was demonstrated (Table 5, Figure 6). The increase in the total height of the roughness profile indicates the deposition of the biofilm not only in the cavities of the material, but also in the creation of new peaks (Figure 7). On the basis of total height of the roughness profile, the thickness of the biofilm layer can be estimated to be around 50 μm (the difference between the total height of the roughness profile of the biofilm material and the reference material).

## 5. Conclusions

Water flowing into the installation did not meet criteria of biological stability. The availability of nutritional substances and an increase in temperature during water residence in the installation resulted in the increase in the number of microorganisms in water and the biofilm formation on internal surfaces of pipes. An increase in turbidity (an average of 1.01 NTU) and decrease in the concentration of chlorine (an average of 74% total chlorine and 63% free chlorine) in water leaving the installation indicated the increase in the risk of the loss of water biological stability and secondary contamination danger.

The development of effective treatment technology, allowing the maintenance of the stability of tap water, and proper selection of installation materials is the key issue in the context of ensuring the health security of consumers. This analysis of obtained results of microbiological tests (ATP, flow cytometry, HTC methods) confirmed that regardless of the material from which the water supply system is made, it is still at risk of formation of biofilm.

Galvanized steel and PE were most susceptible to microorganism adhesion, while PVC and chromium-nickel steel created the least suitable conditions for the development of biofilm (galvanized steel > PE > chromium-nickel steel > PVC).

The most similar amounts of microorganisms occurred on galvanized steel and PE. The biggest differences in the settlement of the surface of materials by microorganisms were found between galvanized steel and PVC. Such a dependence was obtained for all the methods used for the quantitative determination of biofilms. However, due to the high dependence of microbiological determinations on external factors, these conclusions should be confirmed with time-consuming and primary statistical research (ongoing).

**Author Contributions:** Conceptualization, D.P., A.W. and A.P.; methodology, D.P., B.T-C., A.W., J.K, J.Ż and A.D.; validation, D.P., A.D. and J.K.; formal analysis, D.P., A.W. and A.D.; writing—original draft preparation, A.W. and D.P.; writing—review and editing, A.W., D.P., A.D. and J.K.; visualization, A.W., J.Ż. and A.D.; supervision, D.P., B.T-C. and J.K.; project administration, D.P.; funding acquisition, D.P., B.T-C., A and J.K.

**Funding:** This research was funded by subsidies for statutory activity (number: DS.BO.17.001).

**Conflicts of Interest:** The authors declare no conflicts of interest.

## References

1. Inkinen, J.; Kaunisto, T.; Pursiainen, A.; Miettinen, I.T.; Kusnetsov, J.; Riihinen, K.; Keinänen-Toivola, M.M. Drinking water quality and formation of biofilms in an office building during its first year of operation, a full scale study. *Water Res.* **2014**, *49*, 83–91. [CrossRef] [PubMed]
2. Azara, A.; Castiglia, P.; Piana, A.; Masia, M.D.; Palmieri, A.; Arru, B.; Maida, G.; Dettori, M. Derogation from drinking water quality standards in Italy according to the European Directive 98/83/EC and the Legislative Decree 31/2001-a look at a recent past. *Ann. Ig.* **2018**, *30*, 517–526. [PubMed]
3. European Commission. *Proposal for a DIRECTIVE OF THE EUROPEAN PARLIAMENT AND OF THE COUNCIL on the Quality of Water INTENDED for Human Consumption (Recast)*; COM/2017/0753 final-2017/0332 (COD); European Commission: Brussels, Belgium, 2017.
4. Zacheus, O.M.; Iivanainen, E.K.; Nissinen, T.K.; Lehtola, M.J.; Martikainen, P.J. Bacterial biofilm formation on polyvinyl chloride, polyethylene and stainless steel exposed to ozonated water. *Water Res.* **2000**, *34*, 63–70. [CrossRef]
5. Lehtola, M.J.; Miettinen, I.T.; Keinänen, M.M.; Kekki, T.K.; Laine, O.; Hirvonen, A.; Vartiainen, T.; Martikainen, P.J. Microbiology, chemistry and biofilm development in a pilot drinking water distribution system with copper and plastic pipes. *Water Res.* **2004**, *38*, 3769–3779. [CrossRef] [PubMed]
6. Lehtola, M.J.; Laxander, M.; Miettinen, I.T.; Hirvonen, A.; Vartiainen, T.; Martikainen, P.J. The effects of changing water flow velocity on the formation of biofilms and water quality in pilot distribution system consisting of copper or polyethylene pipes. *Water Res.* **2006**, *40*, 2151–2160. [CrossRef] [PubMed]
7. Niquette, P. Impacts of pipe materials on densities of fixed bacterial biomass in a drinking water distribution system. *Water Res.* **2000**, *34*, 1952–1956. [CrossRef]
8. Młyńska, A.; Zielina, M. A comparative study of portland cements CEM I used for water pipe renovation in terms of pollutants leaching from cement coatings and their impact on water quality. *J. Water Supply Res. Technol.-Aqua* **2018**, *67*, 685–696.
9. Manuel, C.M.; Nunes, O.C.; Melo, L.F. Dynamics of drinking water biofilm in flow/non-flow conditions. *Water Res.* **2007**, *41*, 551–562. [CrossRef]
10. Papciak, D.; Tchórzewska-Cieslak, B.; Pietrucha-Urbanik, K.; Pietrzyk, A. Analysis of the biological stability of tap water on the basis of risk analysis and parameters limiting the secondary growth of microorganisms in water distribution systems. *Desalin. Water Treat.* **2018**, *117*, 1–8. [CrossRef]
11. Wolska, M.; Mołczan, M. Assessment of the Stability of Water Entering the Water Supply Network. *Environ. Prot.* **2015**, *37*, 51–56. Available online: http://www.os.not.pl/docs/czasopismo/2015/4-2015/Wolska_4-2015.pdf (accessed on 7 May 2019). (in Polish).
12. Rabin, R. The Lead Industry and Lead Water Pipes "A MODEST CAMPAIGN.". *Am. J. Public Health* **2008**, *98*, 1584–1592. [CrossRef]
13. Pietrzyk, A.; Papciak, D. The influence of water treatment technology on the process of biofilm formation on the selected installation materials. *J. Civ. Eng. Environ. Archit.* **2017**, *64*, 131–142.

14. Bucheli-Witschel, M.; Kötzsch, S.; Darr, S.; Widler, R.; Egli, T. A new method to assess the influence of migration from polymeric materials on the biostability of drinking water. *Water Res.* **2012**, *46*, 246–260. [CrossRef] [PubMed]
15. Chowdhury, S. Heterotrophic bacteria in drinking water distribution system: A review. *Environ. Monit. Assess.* **2012**, *184*, 6087–6137. [CrossRef] [PubMed]
16. Schwartz, T.; Hoffmann, S.; Obst, U. Formation and bacterial composition of young, natural biofilms obtained from public bank-filtered drinking water systems. *Water Res.* **1998**, *32*, 2787–2797. [CrossRef]
17. Waines, P.L.; Moate, R.; Moody, A.J.; Allen, M.; Bradley, G. The effect of material choice on biofilm formation in a model warm water distribution system. *Biofouling* **2011**, *27*, 1161–1174. [CrossRef] [PubMed]
18. Yu, J.; Kim, D.; Lee, T. Microbial diversity in biofilms on water distribution pipes of different materials. *Water Sci. Technol.* **2010**, *61*, 163–171. [CrossRef] [PubMed]
19. Waller, S.A.; Packman, A.I.; Hausner, M. Comparison of biofilm cell quantification methods for drinking water distribution systems. *J. Microbiol. Methods* **2018**, *144*, 8–21. [CrossRef] [PubMed]
20. Lehtola, M.J.; Miettinen, I.T.; Lampola, T.; Hirvonen, A.; Vartiainen, T.; Martikainen, P.J. Pipeline materials modify the effectiveness of disinfectants in drinking water distribution systems. *Water Res.* **2005**, *39*, 1962–1971. [CrossRef] [PubMed]
21. Mandelbrot, B.B. *Fractals: Form, Chance and Dimension*; Freeman: San Francisco, CA, USA, 1977.
22. Konkol, J.; Prokopski, G. The necessary number of profile lines for the analysis of concrete fracture surfaces. *Struct. Eng. Mech.* **2007**, *25*, 565–576. [CrossRef]
23. Wolska, M. Efficiency of removal of biogenic substances from water in the process of biofiltration. *Desalin. Water Treat.* **2015**, *57*, 1354–1360. [CrossRef]
24. Obiri-Danso, K.; Weobong, C.A.A.; Jones, K. Aspects of health-related microbiology of the Subin, an urban river in Kumasi, Ghana. *J. Water Health* **2005**, *3*, 69–76. [CrossRef] [PubMed]
25. Djuikom, E.; Njine, T.; Nola, M.; Sikati, V.; Jugnia, L.B. Microbiological water quality of the Mfoundi River watershed at Yaoundé, Cameroon, as inferred from indicator bacteria of fecal contamination. *Environ. Monit. Assess.* **2006**, *122*, 171–183. [CrossRef] [PubMed]
26. Donlan, R.M. Biofilms: Microbial Life on Surfaces. *Emerg. Infect. Dis.* **2002**, *8*, 881–890. [CrossRef] [PubMed]
27. Straub, T.M.; Gerba, C.P.; Zhou, X.; Price, R.; Yahya, M.T. Synergistic inactivation of Escherichia coli and MS-2 coliphage by chloramine and cupric chloride. *Water Res.* **1995**, *29*, 811–818. [CrossRef]
28. Thurman, R.B.; Gerba, C.P.; Bitton, G. The molecular mechanisms of copper and silver ion disinfection of bacteria and viruses. *Crit. Rev. Environ. Control* **1989**, *18*, 295–315. [CrossRef]
29. Stainless Steel Pipes. Available online: http://www.instsani.pl/606/rury-stalowe-nierdzewne (accessed on 14 April 2019).
30. Pietrucha-Urbanik, K.; Tchórzewska-Cieślak, B.; Papciak, D.; Skrzypczak, I. Analysis of chemical stability of tap water in terms of required level of technological safety. *Arch. Environ. Prot.* **2017**, *43*, 3–12. [CrossRef]
31. Nawrocki, J.; Raczyk-Stanisławiak, U.; Świetlik, J.; Olejnik, A.; Sroka, M.J. Corrosion in a distribution system: Steady water and its composition. *Water Res.* **2010**, *44*, 1863–1872. [CrossRef]
32. Zamorska, J. Biological Stability of Water after the Biofiltration Process. *J. Ecol. Eng.* **2018**, *19*, 234–239. [CrossRef]
33. Lu, C.; Chu, C. Effects of Acetic Acid on the Regrowth of Heterotrophic Bacteria in the Drinking Water Distribution System. *World J. Microbiol. Biotechnol.* **2005**, *21*, 989–998. [CrossRef]
34. Lehtola, M. Microbially available organic carbon, phosphorus, and microbial growth in ozonated drinking water. *Water Res.* **2001**, *35*, 1635–1640. [CrossRef]
35. Lautenschlager, K.; Boon, N.; Wang, Y.; Egli, T.; Hammes, F. Overnight stagnation of drinking water in household taps induces microbial growth and changes in community composition. *Water Res.* **2010**, *44*, 4868–4877. [CrossRef] [PubMed]
36. Kooij, D. Potential for biofilm development in drinking water distribution systems. *J. Appl. Microbiol.* **1998**, *85*, 39S–44S. [CrossRef] [PubMed]

37. Augustynowicz, J.; Niereniński, M.; Jóźwiak, A.; Prędecka, A.; Russel, S. The influence of basic physicochemical parameters on the number of psychrophilic and mesophilic bacteria in the Vistula River waters. *Water-Environ.-Rural Areas* **2017**, *17*, 5–13. Available online: file:///C:/Users/HP/Downloads/Augustynowicz%20i%20in%20(1).pdf (accessed on 7 May 2019). (in Polish).

38. Hallam, N.; West, J.; Forster, C.; Simms, J. The potential for biofilm growth in water distribution systems. *Water Res.* **2001**, *35*, 4063–4071. [CrossRef]

39. Gillespie, S.; Lipphaus, P.; Green, J.; Parsons, S.; Weir, P.; Juskowiak, K.; Jefferson, B.; Jarvis, P.; Nocker, A. Assessing microbiological water quality in drinking water distribution systems with disinfectant residual using flow cytometry. *Water Res.* **2014**, *65*, 224–234. [CrossRef] [PubMed]

40. Francisque, A.; Rodriguez, M.J.; Miranda-Moreno, L.F.; Sadiq, R.; Proulx, F. Modeling of heterotrophic bacteria counts in a water distribution system. *Water Res.* **2009**, *43*, 1075–1087. [CrossRef] [PubMed]

*Article*

# Using Nodal Infection Risks to Guide Interventions Following Accidental Intrusion due to Sustained Low Pressure Events in a Drinking Water Distribution System

**Fatemeh Hatam** [1,*]**, Mirjam Blokker** [2]**, Marie-Claude Besner** [3]**, Gabrielle Ebacher** [4] **and Michèle Prévost** [1]

[1]  NSERC Industrial Chair in Drinking Water, Department of Civil, Geological and Mining Engineering, Polytechnique Montréal, CP 6079, Succ. Centre-ville, Montréal, QC H3C 3A7, Canada
[2]  KWR Watercycle Research Institute, Groningenhaven 7, 3433 PE Nieuwegein, The Netherlands
[3]  R&D Engineer, Water Service, City of Montreal, Montréal, QC H3C 6W2, Canada
[4]  Technical Engineer, Environment Service, Drinking Water Division, City of Laval, QC H7V 3Z4, Canada
*  Correspondence: fatemeh.hatam@polymtl.ca; Tel.: +1-514-340-4711 (ext. 2983)

Received: 10 June 2019; Accepted: 28 June 2019; Published: 3 July 2019

**Abstract:** Improving the risk models to include the possible infection risk linked to pathogen intrusion into distribution systems during pressure-deficient conditions (PDCs) is essential. The objective of the present study was to assess the public health impact of accidental intrusion through leakage points in a full-scale water distribution system by coupling a quantitative microbial risk assessment (QMRA) model with water quality calculations based on pressure-driven hydraulic analysis. The impacts on the infection risk of different concentrations of *Cryptosporidium* in raw sewage (minimum, geometric mean, mean, and maximum) and various durations of intrusion/PDCs (24 h, 10 h, and 1 h) were investigated. For each scenario, 200 runs of Monte Carlo simulations were carried out to assess the uncertainty associated with the consumers' behavioral variability. By increasing the concentrations of *Cryptosporidium* in raw sewage from 1 to 560 oocysts/L for a 24-h intrusion, or by increasing the duration of intrusion from 1 to 24 h, with a constant concentration (560 oocysts/L), the simulated number of infected people was increased by 235-fold and 17-fold, respectively. On the first day of the 1-h PDCs/intrusion scenario, a 65% decrease in the number of infected people was observed when supposing no drinking water withdrawals during low-pressure conditions at nodes with low demand available (<5%) compared to no demand. Besides assessing the event risk for an intrusion scenario, defined as four days of observation, the daily number of infected people and nodal risk were also modeled on different days, including during and after intrusion days. The results indicate that, for the case of a 1-h intrusion, delaying the start of the necessary preventive/corrective actions for 5 h after the beginning of the intrusion may result in the infection of up to 71 people.

**Keywords:** QMRA; sustained pressure drops; accidental intrusion; infection risk from *Cryptosporidium*; pressure-driven hydraulic analysis

## 1. Introduction

Distribution system (DS) deficiencies may play a role in the occurrence of waterborne disease outbreaks [1]. Ageing of pipeline infrastructure is going to become more problematic over time by increasing the probability of experiencing sustained low/negative pressure conditions in the network (pipe breaks), leading to possible intrusion from points of leakage. Assessment of public health risk associated with such type of events may be achieved through modeling. While reliable hydraulic and water quality models can be used to simulate ingress of contaminated water and its propagation into a network, the use of quantitative microbial risk assessment (QMRA) models is required to estimate the potential health risk. QMRA and management approaches can contribute in bringing safer water to consumers [2].

**Modeling of water quality under pressure deficient conditions.** Integration of pressure-driven hydraulic analysis into QMRA models is required for a more accurate risk analysis of water contamination resulting from accidental intrusion under sustained pressure-deficient conditions (PDCs). In such conditions, a reliable estimation of intrusion points, contamination mass rate entering the DS, and fate/transport of contamination through the network cannot be achieved using traditional demand driven-analysis (DDA) models such as EPANET 2 [3]. Pressure-driven analysis (PDA) was coupled to single species water quality modeling to optimize management strategies (e.g., flushing and isolation actions) by minimizing the mass of consumed contaminant [4–6]. A more detailed literature review on hydraulic and water quality modeling under sustained PDCs can be found elsewhere [7].

**Applications of QMRA to drinking water DSs.** Despite evidence of drinking water DS deficiencies causing infectious waterborne diseases [8,9], the majority of QMRA work has been devoted to assessing risk of drinking water treatment failures [2]. Viñas et al. [10] and Hamouda et al. [11] presented detailed literature reviews on QMRA models applied to microbial contaminants in drinking water DSs. Besner et al. [12] developed a conceptual model to assess the public health risk associated with intrusion events. QMRA models have been applied to real DSs to evaluate the infection risk associated with the presence of viruses resulting from intrusion events caused from transient PDCs [13–15]. Standard QMRA models consider the water is consumed randomly at any time or at fixed times during the day [14,16,17]. The timing of water withdrawals for drinking purpose is an important factor when assessing the probability of infection as a result of intrusion events and may not be the same as the timing of the total consumption [17,18]. An improved QMRA that integrates the consumer's behavior (probability density functions (PDFs) of the numbers of glasses and the volume consumed, and kitchen tap use) was developed and applied to assess the infection risk associated with contamination after main repairs [18,19]. They investigated the impact of different parameters such as the location of contamination and the times of valve openings on the infection risk with various pathogens (*Campylobacter*, *Cryptosporidium*, *Giardia* and rotavirus), in the absence of any disinfectant residual. Schijven et al. [20] also considered consumer behavior to estimate the infection risk from ingestion of contaminated water or inhalation of contaminated aerosol droplets in the case of intentional contamination of different durations and seeding concentrations in a DS.

Improving estimations of the infection risks due to sustained pressure deficient conditions requires numerical approaches that produce realistic estimations of nodal ingress volumes, predictions of propagation throughout the network, and integration of the consumer's behavior during and after pressure losses. Besner et al. [16] emphasized the necessity of performing PDA instead of DDA to simulate the infection risk associated with PDCs in future studies. Besides low pressure, the presence of external contamination and pathways are essential for intrusion to occur [21]. Adjusting the presence of potential pathway for intrusion based on the state of decay of the piping has been proposed [22,23].

The primary objective of this work was to estimate the infection risk associated with accidental intrusion through leakage points into a DS as a result of unplanned sustained low/negative pressure events (24 h, 10 h, and 1 h). To achieve this goal, several original improvements to the various models were made. First, the QMRA model developed by Blokker et al. [18] was customized and linked with water quality calculations based on a pressure-driven hydraulic analysis. Then, the estimated contamination mass rate at each intrusion node was adjusted by the assigned leakage demand (proxy for pipe age and material) and the pressure values during PDCs, computed using PDA. Finally, to better simulate the consumers behavior during low-pressure conditions, the consumption of tap water was adjusted based on demand availability (no demand or <5%) on the infection risk. The secondary objective of this work was to propose a basis for the analysis of risk to guide the definition of areas subjected to a boil water advisory or corrective actions. To achieve this goal, we assessed the potential use of the temporal (daily versus event) and spatial distribution of nodal risks to determine the location and the duration of advisories. To the knowledge of the authors, no study so far has quantified the infection risk of accidental intrusion resulting from sustained PDCs, using realistic PDA to adjust intrusion volume for nodal pressure, perform water quality analysis and integrate the impact of demand availability on the consumption during pressure drops.

## 2. Methodology

The QMRA model developed by Blokker et al. [18] was customized to be coupled with water quality calculations based on pressure-driven hydraulic analysis. The model was used to quantify the infection risk associated with accidental intrusion events as a result of sustained PDCs in a full-scale DS. The main steps for risk analysis are exposure analysis and calculation of infection risk. A simplified flow chart of the QMRA steps is illustrated in Figure 1. These steps include: (a) simulating the hydraulic behavior of the network under the intended PDCs to define the intrusion nodes, intrusion flow rates (based on size of opening leaks and pressure differential), and nodes with unsatisfied demand; (b) defining the outside pipe conditions to calculate the potential contaminant mass rate entering the system; (c) modeling fate/transport of ingress microorganisms through network; (d) specifying the microbial exposure (dose) considering consumers' drinking water behavior; and (e) estimating the risk of infection based on dose–response models.

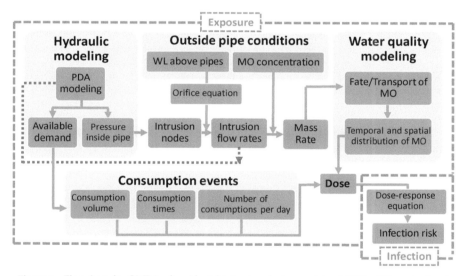

**Figure 1.** Flowchart for QMRA of accidental intrusion during sustained PDCs; WL, water level; MO, microorganism.

### 2.1. Exposure Analysis

#### 2.1.1. Hydraulic and Water Quality Analysis

To estimate the ingested dose, fate/transport of contaminants through the network should first be estimated using appropriate hydraulic and water quality models. Water quality modeling based on PDA was performed using WaterGEMS V8i (SELECTseries 5) [24]. Transport of *Cryptosporidium* oocysts through the network was simulated over time and, because *Cryptosporidium* is highly resistant to chlorine disinfection [25], the chlorine decay was not included in the model. Sewage is defined as the source of contamination outside the pipes. Minimum, geometric mean, arithmetic mean, and maximum levels of *Cryptosporidium* in sewage were assumed to be 1, 6, 26, and 560 oocysts/L, respectively [26].

The DS model used in this study includes 30,077 nodes and 3 water treatment plants (WTPs), which serve nearly 400,000 residents. More details on the simulated full-scale network can be found in Hatam et al. [7]. The unplanned shutdown of one WTP was simulated and a 5 m decrement in the outlet pressure of the two other WTPs was assumed as a result of the flow-rate increase. It should be noted that the two other WTPs might (partially) compensate the shutdown of the other WTP as the entire network is hydraulically interconnected. Following the shutdown duration (1, 10 or 24 h), the simulation was continued for 3 days to investigate the long-term public health impacts of the accidental intrusion events in this large DS. The impacts of intrusion duration on exposure and, consequently, risk of infection were studied. More details on accidental intrusion modeling can be found in the Supplementary Materials. Nodes with pressure head less than 1 m were considered as the potential intrusion sites (Figure S4). In the hydraulic model, for the sake of simplicity, the demand is considered constant during the day and equal to the peak hour demand (i.e., 19:00) for the scenarios of 1, 10 and 24 h of PDCs/intrusion. Additional scenario with the daily water consumption pattern in the hydraulic model was studied for the intrusion event resulting from 1 h PDCs set to start at 18:30.

2.1.2. Consumption Events

The temporal concentrations of *Cryptosporidium* calculated from water quality analysis were then imported into MATLAB (MathWorks, Natick, MA, USA) where the QMRA was performed for exposure assessment and dose–response analysis. Consumption events or consumers' behavior in this study refer to: (1) the volume of consumption; (2) the number of times that one fills a glass; and (3) the times at which the glass is filled from the tap. In the present study, consumption times corresponded to the water use at the kitchen tap as proposed by Blokker et al. [18]. In the simulations, the average kitchen tap use was then modified for each node of the studied network based on the nodal residential demand and the availability of demand, calculated from PDA under PDCs. In this study, the average kitchen tap use for non-residential nodes (about 60% of the nodes) was set to zero. This differed from Blokker et al. [18], who adjusted the average kitchen tap use at certain times to include zero demand periods identified by detailed residential demand. In this study, to account for demand satisfaction as computed by PDA at each node, the kitchen tap use was set to zero at times when there was no demand available under PDCs (Figure S1). For PDCs with some demand satisfaction, it was assumed that consumers can adjust the filling duration based on the available flow at the tap. If the PDCs did not last for the whole day, the total daily volume of water consumed by each person at the nodes with no demand under PDCs would not be affected. The sensitivity of the results to the demand satisfaction ratio (DSR) was investigated in an additional scenario by fixing the kitchen tap use to zero at the time when there is low (<5%) demand available at the nodes. This approach is more realistic as the required time to fill a glass of water at a kitchen tap will increase by more than 20 times when the DSR is less than 5%.

The other important parameter for estimating the risk of exposure to microbial contamination is the volume of water that is ingested per person per day. The number of times each person would fill his/her glass or bottle during a day was estimated using a Poisson distribution. The ingested volume at each filling time was defined by a lognormal distribution. Due to the lack of information for the studied network, the data from Blokker et al. [18] were used for the simulation and more details can be found in their paper.

In this study, the hydraulic and water quality conditions were assumed to be known for each scenario, and 200 runs of Monte Carlo simulations were performed to investigate consumers' behavior. In each Monte Carlo run, the number and times of consumption events as well as the ingested volume for each consumption event were randomly picked for each person every day of the simulation.

In the studied hydraulic model, the total nodal demands could be a combination of different types of demand defined as: residential, commercial, industrial, institutional, municipal or, leakage. In total, 11,194 of the nodes included residential demand. To determine the number of people supplied per node, the residential demand per node was considered and the daily per capita average demand was set to 220 L/person/day. Consequently, only the residential exposure from tap water as a result of the simulated accidental intrusion was investigated (e.g., exposure at school was not considered). More information on the estimation of the number of people at each node and the distribution of population is in the Supplementary Materials. Dose is equal to the number of consumed pathogens and was calculated by multiplying the intake volume by the concentration of pathogens at the time of withdrawal. This step was repeated for all the glasses that a person takes over the simulation duration, which is 1 day for daily risk and 4 days for the event risk. For each person, the total dose was calculated by summing the dose in each glass consumed.

*2.2. Calculation of Infection Risk*

Dose–response analysis was performed to calculate the infection risk for each person resulting from accidental intrusion during sustained PDCs. The computed dose was implemented in the dose–response model employed by Blokker et al. [19] for *Cryptosporidium* using the median (50th percentile) and maximum (100th percentile) dose–response relationships. The median infection risk is reported everywhere in this study unless otherwise stated.

The calculated infection risks of all the people in the network were summed up and rounded to the nearest integer greater than or equal to the calculated value to estimate the equivalent number of infected people for the simulated event [18]. The number of infected people was calculated either for the whole observation period (4 days) or for each day separately. To calculate the nodal risk, the infection risks corresponding to all the people at the same node were summed up.

## 3. Results

**Estimating ingress volumes.** Histograms of nodal pressures and demand satisfaction ratios (DSRs: available nodal demand divided by the required demand) using PDA are illustrated in Figure 2. Fewer than 1% of the nodes (93 nodes) were prone to intrusion as they experienced pressures less than 1 m under PDCs, which corresponded to the set pressure head above pipes. For about 30% of the nodes, the pressure was less than or equal to the required pressure value assumed in this study for full demand satisfaction (15 m). The DSRs for these nodes are shown in Figure 2b, excluding nodes with no required demand. Figure 2b shows that 1103 nodes have a DSR of less than 50% during depressurization.

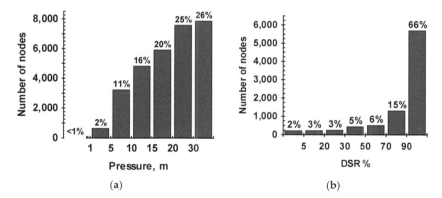

**Figure 2.** Distribution of: (**a**) nodal pressures for the whole network (30,077 nodes); and (**b**) demand satisfaction ratios (DSRs) for nodes under pressure-deficient conditions (8578 nodes), excluding the nodes with zero demand.

The distribution of intrusion flow rates at the ingress nodes is illustrated in Figure 3. The maximum flow rate was 56 L/h and about half of the nodes had an intrusion flow rate less than 5 L/h. The contaminated water entered the network at a flow rate of 804 L/h through all the leakage orifices.

For the scenarios of 10 and 24 h PDCs, the intrusion flow rate at each node remained constant during the event because of the use of a constant demand. As the 1 h event, with daily consumption pattern, was assumed to occur at the peak demand hour, the nodal intrusion flow rates also corresponded to those shown in Figure 3.

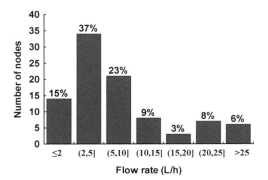

**Figure 3.** Distribution of nodal intrusion flow rates through 93 leak openings under the simulated pressure-deficient conditions.

**Concentrations of pathogens in sewage.** To cover different consumption behaviors, 200 Monte Carlo simulations were carried out for each scenario of *Cryptosporidium* concentration in sewage (1, 6, 26, and 560 oocysts/L). The resulting cumulative probability distributions of the number of infected people are plotted in Figure 4. In this figure, the solid lines correspond to the median infection risk, and the dotted lines are the maximum infection risk. For all concentrations, the number of infected people associated to the maximum infection risk was increased by about two folds compared to the median infection risk. For the concentration of 560 oocysts/L, 50% of the consumption events led to at least 1378 (2652) infected people considering the median (maximum) infection risk. As expected, the number of infected people increased when the *Cryptosporidium* concentration increases from 1 to 560 oocysts/L.

**Consumption behavior.** Figure 5 shows the sensitivity of the number of infected people over the four-day observation period to the volume of consumption (300 mL, 500 mL or 1 L per day per person) and number of glasses per day (1, 3, or 10). A total of nine scenarios were considered with a *Cryptosporidium* concentration of 560 oocysts/L and 24 h of PDCs. As expected, lower volumes of unboiled tap drinking water per person per day largely reduced the infection risk. By decreasing the volume by half (500 mL), the number of infected people decreased by 40%; decreasing the volume to 300 mL reduced the risk further by about 60%. By increasing the number of glasses per day from 1 to 3, 19 more people were likely to be infected for a 300 mL volume, and this value became 62 for a 1 L consumption volume per day per person (based on the values of $F(x) = 1$).

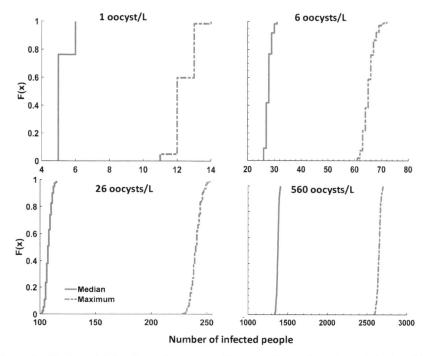

**Figure 4.** Number of infected people corresponding to median and maximum infection risks resulting from a 24-h depressurization; 200 Monte Carlo simulations (consumption events) for each *Cryptosporidium* concentration: 1, 6, 26, and 560 oocysts/L; number of infected people corresponds to the cumulative dose over four days of observation; F(x): probability that the median/maximum number of infected people will be less than or equal to x.

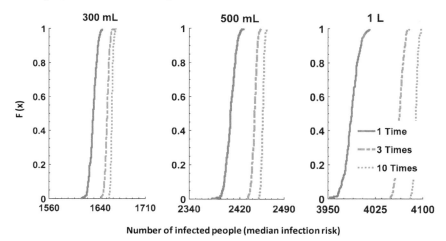

**Figure 5.** Impact of consumption volumes and number of glasses per day on the number of infected people corresponding to median infection risk over a four day-period; *Cryptosporidium* concentration = 560 oocysts/L; the x-axis scale is the same between the plots (150 people).

**Duration.** Shorter duration PDCs can take place in real networks because of WTP shutdowns, pipe breaks or fire flows. The cumulative probability distribution of the number of infected people for 200 random consumption behaviors is shown for different durations of PDCs: 1, 10, and 24 h (Figure 6). In all scenarios, the timing of the event was adjusted so that the network experienced low/negative pressures at the peak consumption time (i.e., 19:00) of the first day. A significant dependence of the infection risk with the intrusion duration was observed: a lower maximum number of infected people (84) was observed for a 1-h intrusion compared to 502 and 1410 for 10 and 24 h intrusion events, respectively.

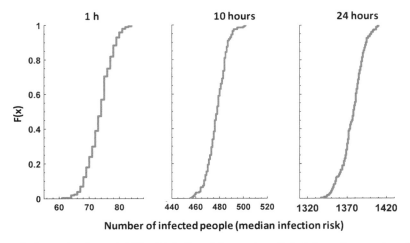

**Figure 6.** Comparing the probability distribution of the number of infected people over a four-day period for 200 Monte Carlo simulations for each duration of PDCs: 1, 10, and 24 h; *Cryptosporidium* concentration in sewage = 560 oocysts/L.

**Spatial distribution of nodal infection risk.** Besides the number of infected people under PDCs, the temporal and geographical distribution of infection risk is also essential in defining appropriate preventive/corrective actions. In this regard, the probability of infection of the individuals who were assigned to the same node were summed up to predict the nodal risk. Figure 7 shows the spatial distribution of risk for above-mentioned scenarios corresponding to the consumption events with the maximum number of infected people ($F(x) = 1$ in Figure 6). As shown, with increasing duration of intrusion event, not only the nodal risks are were, but also larger areas were at risk.

**Daily risk for the 1-h event with daily demand patterns.** For the prior analyses, demand was considered constant during the day and equal to the peak hour demand (i.e., 19:00) in the hydraulic model. The reason is that adjusting different intrusion volumes and nodes at each hour of the duration of PDCs using PDA would be computationally intensive. However, we investigated a 1 h PDCs/intrusion using the daily water consumption pattern in the hydraulic model to assess its impact on the infection risk. Over four days of observation, the maximum number of infected people increased to 99 (Figure S3) with demand patterns compared to 84 with a constant demand in the hydraulic model (Figure 6, 1 h).

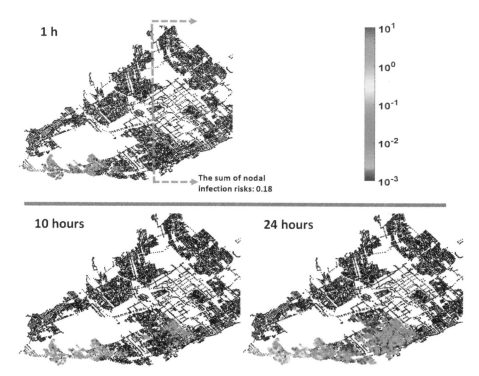

**1 h**

The sum of nodal infection risks: 0.18

$10^1$

$10^0$

$10^{-1}$

$10^{-2}$

$10^{-3}$

**10 hours**

**24 hours**

**Figure 7.** Spatial distribution of nodal risks for three durations of PDCs: 1, 10, and 24 h; *Cryptosporidium* concentration in sewage = 560 oocysts/L; nodes with an infection risk below $1 \times 10^{-3}$ are drawn in black; infection risks corresponding to consumption events with F(x) = 1 (Figure 6) are illustrated.

Figure 8 illustrates the daily probability of the number of people infected by *Cryptosporidium* according to different consumption behaviors for the day that intrusion occurred (at 18:30) and the three days post-intrusion. The day after the event, the maximum number of infected people was reduced by 59% as compared to the event day. It indicates that, over time, the contaminated water left the network as large volumes of water were used for purposes other than drinking, such as toilet flushing and industrial usage. The maximum numbers of infected people for Days 1–4 were 71, 29, 3 and 1, respectively.

For Days 1–4, the total nodal risk corresponding to the consumption event with the maximum number of infected people (F(x) = 1 in Figure 8) was estimated, and the spatial distribution is plotted in Figure 9. The number of nodes at high risk decreased from Day 1 to Day 4 as well as the extent of the areas at risk. At the end of the first day, when the intrusion ended, the nodal infection was $\leq 1 \times 10^{-7}$ at 29,754 nodes and higher than $1 \times 10^{-4}$ at 123 nodes. Only 16 of the nodes showed total nodal risks equivalent to more than one person. On Day 2, the total number of infected people through the whole network decreases to 29 compared to 71 for Day 1, but the number of nodes with an infection risk $\leq 1 \times 10^{-7}$ was lower compared to Day 1. The reason is that *Cryptosporidium* oocysts reached more nodes in the network on Day 2, but at lower concentrations as the ingress volume became diluted and flushed out. On Day 2, the nodal infection risk was more than one only at four nodes. On Days 3 and 4, the nodal infection risk was below one for all the nodes.

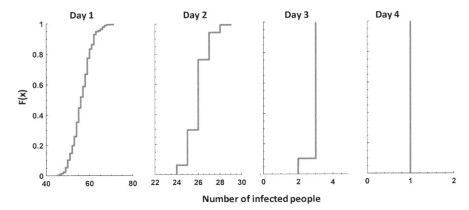

**Figure 8.** Number of infected people corresponding to median infection risk for Days 1–4 for the scenario of 1 h of PDCs with daily consumption patterns; $C_{out}$ = 560 oocysts/L; 200 Monte Carlo simulations (consumption events) every day.

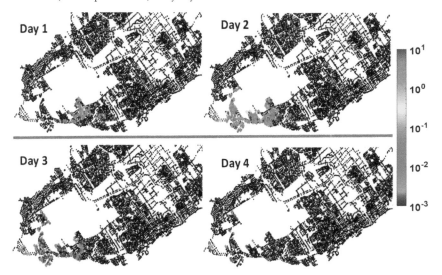

**Figure 9.** Spatial distribution of nodal risk; Days 1–4 for the scenario of 1 h of PDCs with daily consumption patterns; $C_{out}$ = 560 oocysts/L; nodes with infection risk below $1 \times 10^{-3}$ are drawn in black; infection risks corresponding to consumption events with F(x) = 1 (Figure 8) are illustrated.

**Impact of demand satisfaction ratio on risk.** In all simulations, when the DSR (pressure ≤ 0) became zero at a node, the kitchen tap use was set to zero. To study the influence of the DSR (shown in Figure 2b) on the risk, the situation where no consumption happened at nodes with a DSR less than 5% was also modeled (Figure 10). For this investigation, the number of infected people following a 1-h PDCs/intrusion was computed on the day that intrusion occurred. As expected, the number of infected people decreased when the consumption only occurred at the nodes with a DSR ≥ 5% during low/negative pressure conditions (Figure 10).

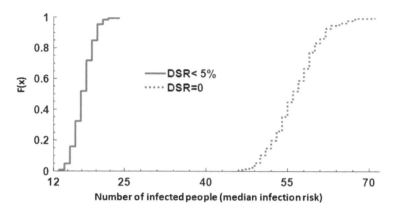

**Figure 10.** Probability distributions of the number of infected people during the first day of simulation when people with a DSR null and less than 5% do not drink water from tap; 200 Monte Carlo simulations for each scenario; $C_{out}$ = 560 oocysts/L with 1 h of PDCs with daily consumption patterns.

## 4. Discussion

**Impact of event duration on the spatial distribution of risk in the network.** During an intrusion event, the intrusion risk was determined by several factors such as the intrusion volume, pathogen concentration, network hydraulics, fate and transport of the contaminants and consumers' behavior. The volume of contaminated water entering the network is a function of the duration of the event. For the events with 1, 10 and 24 h of sustained depressurization, the estimated intrusion volumes through all leak openings were 0.8, 8 and 19 m$^3$, respectively. Using the orifice equation, some studies have produced estimates of the intrusion volumes through leakage points for transient PDCs [1,13,22]. The total intrusion volumes resulting from a momentary pump shutdown for different intrusion conditions through leakage orifices and submerged air vacuum valves (AVVs) ranged from 10 to 360 L in the same network [22]. In contrast, these authors also showed that the maximum volume entering through a single submerged AVV during a transient could be about 95 times larger than the maximum volume entering through a single leakage orifice (227 L versus 2.4 L). In their study, the modeled intrusion volume was driven by the global leakage rate (5% versus 40%) and pressure differential. However, as these authors also stated, the orifice size at a given node should reflect the local leakage demand. Using Monte Carlo simulations, Gibson et al. [23] investigated the impact of head differences, diameter of orifices, pipe age (number of holes), and low pressure duration on the intrusion volumes during transient negative pressure events. For a 25-year-old pipe, the probability of an intrusion volume greater than 10 L was low (1%), while it increased to 70% for a 150-year-old pipe.

In the current study, the orifice size at each node was considered proportional to the assigned nodal leakage demand in the calibrated model under normal operating conditions as described in detail by Hatam et al. [27]. In the test DS, leakage demand reflects the state of pipes; older areas with aging cast iron being the dominant pipe material has higher leakage and thus offers more potential entry points for contaminated water. In this study, the effect of soil–leak interactions was ignored and the exponent in the orifice equation was considered equal to the theoretical value (0.5) that is valid for fixed leak openings. It was confirmed that the variation of the area of round hole with pressure is negligible and therefore the leakage exponent was close to 0.5 [28,29]. However, for longitudinal slits that have large head-area slope, a modified orifice equation should be used in which the leakage exponent can change within 0.5 to 1.5 [30].

In this study, long durations of PDCs were considered as opposed to relatively short durations of low and negative pressures. Sustained PDCs are reported in the literature due to transmission main repairs [12,31] and can happen during power outages. This type of event may be less frequent than transient pressure fluctuations, but of graver consequences, as shown by the potentially larger intrusion volumes. The duration of transient negative or low pressures is a key factor affecting the virus infection risks estimated by QMRA [13–15]. As expected, for the simulated sustained PDCs, the number of infected people for the three different intrusion durations showed strong dependency on the intrusion duration (Figure 6), as it determines the total amount of *Cryptosporidium* oocysts introduced into the network. The maximum number of infected people was reduced to less than half when the intrusion duration decreased from 24 h (1410) to 10 h (502), and even more so if the event only lasted 1 h (84). Our results are in agreement with those of Schijven et al. [20], who used QMRA to investigate the impact of intentional contamination. Exposed persons were increased by 2–3 folds when the duration of the injection of contaminants increased from 10 to 120 min.

More importantly, in this study, we showed that the duration determined the areas with high pathogen concentrations corresponding to a potentially significant infection risk. The geographical distribution of the nodal risk shown in Figure 7 emphasizes the importance of considering the duration of PDCs/intrusion when issuing sectorial boil water advisories (BWA) as well as other preventive/corrective actions. For 24 and 10 h intrusion events, the zones at risk were more or less the same with different risk levels. However, for a much shorter duration of intrusion (1 h), the zones at elevated risk were significantly reduced (Figure 8). The arbitrary cutoff line in Figure 8 can be used to compare the summation of the total risks at nodes in different zones affected by contaminated ingress water. On its right side, a very small cumulative risk of 0.2 infection for the 1 h intrusion was observed; this risk increased to 1.4 and 3.5 for the intrusion events of 10 and 24 h, respectively. These values include all low nodal risks ($\leq 1 \times 10^{-3}$), which are not plotted in Figure 7 for clarity.

**Concentration of *Cryptosporidium* in ingress water.** There are scarce data on the actual concentrations of pathogens in ingress water. Concentrations of pathogens in ingress water could range from those found in wastewater, representing a high-risk scenario of ingress directly from undiluted sewage [26], to the much lower concentrations measured in trench water, urban groundwater or runoff [32,33]. The number of infected people increased from 6 to 1410 when *Cryptosporidium* concentrations increased from 1 to 560 oocysts/L (Figure 4, median) for the worst-case consumption event (out of 200) (F(x) = 1). In agreement with our results, the contaminant concentration outside the pipe ranked among the top factors in previous QMRA studies [13,15,18,34]. When using the maximum dose–response relationship rather than the median relationship to account for uncertainties, the maximum number of infected people increased about two folds (Figure 4). The magnitude of differences between the median and maximum dose–response relationships is a critical factor to consider as recent evidence suggests that even higher dose–response values for *C. hominis* should be considered [2,25]. Therefore, both the concentrations and the selection of the dose–response will contribute to uncertainty [2].

**Consumption behavior.** Standard QMRA models usually consider only one consumption event per day [14,15] or a constant volume of consumption per day for every person at fixed hours [16,21]. For the 24 h scenario, the amount of water consumed daily from the kitchen tap had a huge impact on the maximum number of infected people, with decreases of ~40% and 60% when consumption was reduced from a baseline of 1 L/day to 500 mL/day and 300 mL/day, respectively. The model was also sensitive, but to a lesser degree, to the number of glasses per day for a fixed volume (Figure 5). Increasing the number of glasses per day from 1 to 10 increased the overall infection risk (by up to 2%) for the 24-h scenario. This rise was more pronounced for larger consumption volumes (Figure 5). Impact of the number of glasses per day was most noticeable when switching from a single consumption event to 3 or 10 consumption events. Blokker et al. [18] and Van Abel et al. [35] also observed that three ingestion volumes per day result in higher numbers of infected people compared to only one withdrawal of the total volume per day.

Several studies have investigated and integrated probabilistic models to better represent the consumers' behavior into QMRA models, including PDFs of volume of unboiled tap water, number of glasses per day, volume per glass, timing of consumption, and household water usage [17,18,20,36].

Blokker et al. [18] fully integrated consumers' behavior using a Poisson distribution for the number of glasses per person per day and a lognormal distribution for the ingested volume per glass and the kitchen tap use. This model was applied to investigate various scenarios of fecal contamination resulting from DS repairs and the potential for preventive actions to mitigate risks of infection. In this study, we used the Blokker model to investigate accidental intrusion due to sustained low/negative pressure event of various durations, adding 200 simulations to quantify the range of risks corresponding to different consumers' behavior. The differences between the numbers of infected people for minimum (F(x) = 0) and maximum (F(x) = 1) probabilities in Figure 8 reveal the potential impact of consumers' behavior for a specific event. The ranges were widest for the first day (from 71 to 46 people, 35% reduction) compared to the following days. The variations observed were less important in the scenarios of 10 and 24 h (Figure 6). Understanding the uncertainty associated with a combination of plausible behaviors appears important.

**Impact of daily demand.** The diurnal consumption patterns result in variable intrusion volumes and numbers of intrusion nodes during different hours of the day because of the variations in nodal pressure values. In this study, the demand was set to peak hour demand, which could lead to overestimation of intrusion volumes if system pressure was not decreased for night flows. On the other hand, fixed peak water demand overestimated the flushing of contaminants from the network by leakage, commercial, industrial, institutional demands, etc. during periods of low human consumption, resulting in an underestimation of the risk. With the scenario of 1 h PDCs/intrusion which incorporates daily demand patterns in the hydraulic model, it was shown that the underestimation was about 15%, which we consider to be acceptable (Figure 6 compared to Figure S3).

**Integrating demand availability from PDCs.** The novelty of this work lies in the coupling of the PDA and QMRA. Unlike DDA, PDA permits identification of areas with demand shortage, allowing for more realistic estimations of consumption based on water availability at the tap during pressure losses. For example, consuming at a DSR of 5% and less would mean that the filling time would increase by more than 20-fold. As shown on Figure 10, the number of infected people on Day 1 decreased sharply from 71 to 24 (65%) if only consumers at nodes with DSR >5% during low/negative pressures were considered. It should be noted that limitations to consumption only occur during the low-pressure conditions. Furthermore, the extent of these differences depends on the consumption time, and the duration and timing of the event. The results shos that restricting drinking water consumption during periods of low or intermittent flow would greatly reduce risks. Therefore, utilities and health authorities could consider educating people not to consume water during these periods of low flow. Further study is needed to define a minimal DSR criteria based on the amount of reduction in infection risk.

**Implication for risk management.** The nodal risks considered the contaminant transport in the network and the probability of coincidence of passage of contaminants at the tap and consumption. However, the spatial and temporal distribution of total nodal risks also reflected the distribution of the population between nodes (Figures 7 and 9). The areas in which to issue a BWA, and those where corrective actions (e.g., flushing) would be effective, can be determined using nodal risk values in reference to an acceptable risk level.

QMRA models have been used to evaluate the efficacy of different mitigation strategies such as BWAs, flushing, and disinfection for reducing the infection risk after main break repairs/transient pressures [14,18,34]. Yang et al. [34] showed that flushing at >0.9 m/s reduced infection risks by 2–3 logs for norovirus, *E. coli* O157:H7 and *Cryptosporidium*. For viral and bacterial pathogens, disinfection with a CT of at least 100 mg·min/L using free chlorine was required after flushing to decrease the risk below the USEPA yearly microbial risk target value ($1 \times 10^{-4}$) [37]. Issuing a system-wide BWA that decreased by 80% the average number of glasses of unboiled water consumed led to a four-fold reduction in the number of infected people [18].

Estimating the daily risk, instead of the event risk, after an intrusion event can guide risk management decisions. The spatial distribution of risk as shown in Figure 9 is a key factor to define the boundaries and duration of sectorial BWAs. Figures 8 and 9 show the contribution of each day to the total event risk over the four-day period. Notably, for the 1-h intrusion, delaying necessary preventive/corrective actions up to 5 h from the start of the intrusion may result in the infection of up to 71 people. After that 5-h mark, a BWA or other preventive/corrective actions would still offer protection for about 33 additional people (sum over the three following days). The reduced benefit of late interventions on the fourth day was evident with only one equivalent infection prevented. Timely response to sustained PDCs is therefore essential and can be achieved by improving sampling strategies using enhanced numerical model [27] and equipping the DS with multiple online pressure sensors and water quality sensors. The duration of the BWA could be adjusted depending on the corrective actions implemented to meet the acceptable risk level for an event.

Figure 11 offers insights into whether pressure during PDCs can be used to determine areas to target for preventive/corrective actions. Pressure during the PDCs determine the extent of intrusion. However, whether contaminants will travel from low-pressure nodes to higher pressure nodes (based on pressure during PDCs) is determined by water paths during normal and PDCs. This was clearly illustrated by the fact that, for the 1-h PDCs, consumption of tap water at nodes other than negative pressure nodes resulted in 63, 28, 3, and 1 infected people on Days 1–4, respectively. This showed that the benefits of avoiding consumption at negative nodes (based on the pressure values under PDCs) after the PDCs was limited, as these values for the whole network, including negative nodes, were 71, 29, 3 and 1, respectively. Even with a pressure criterion of 15 m, the number of infected people on Day 2 would be significant (6) (Figure 11). These results are consistent with the study by Hatam et al. [27] who showed that *E. coli* can be transported to higher pressure zones (up to ~40 m) in the absence of disinfectant residuals during a 5-h PDCs/intrusion. Our results emphasize that issuing sectorial BWAs based only on pressure is not adequate to protect the population against infection, even for the scenario of 1-h PDCs/intrusion with a high *Cryptosporidium* concentration (560 oocyst/L). The simulation of the fate and transport of contaminants is necessary to define an effective sectorial BWA.

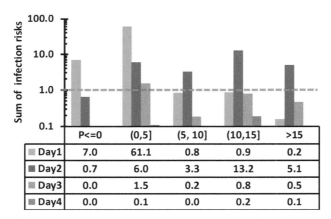

| | P<=0 | (0,5] | (5, 10] | (10,15] | >15 |
|---|---|---|---|---|---|
| Day1 | 7.0 | 61.1 | 0.8 | 0.9 | 0.2 |
| Day2 | 0.7 | 6.0 | 3.3 | 13.2 | 5.1 |
| Day3 | 0.0 | 1.5 | 0.2 | 0.8 | 0.5 |
| Day4 | 0.0 | 0.1 | 0.0 | 0.2 | 0.1 |

**Figure 11.** Number of infected people for different pressure (P) ranges (based on the pressure values under PDCs) on Days 1–4; Infection risks corresponding to the consumption event with F(x) = 1 (Figure 8) are illustrated. The event starts at 18:30 on Day 1 for a duration of 1 h. Daily patterns in the hydraulic model.

In future work, reporting the hourly risk, instead of the daily risk, could be helpful to utilities to define preventive/corrective actions and timely response. In this study, the PDCs occurred at 18:30 on Day 1, therefore some of the daily demands were already satisfied before the intrusion event. The timing of the event impacts the infection risk, which needs to be investigated in future studies. Blokker et al. [18] showed limited effect for timing of repairs.

Although the field validation of the transport of pathogens and indicators appears desirable, it is however not feasible to conduct in complex operating distribution systems. Such validation would require extensive monitoring during intentional extended loss of pressure events and monitoring of infections by an epidemiological investigation that utilities and health authorities will not allow. The conservative modelling presented in this study nevertheless demonstrates the value of numerical tools combined to QMRA to quantify risk and assist utilities and regulators.

## 5. Conclusions

- An approach is proposed to couple QMRA and water quality calculations based on pressure-driven hydraulic analysis to assess the infection risk under sustained low/negative pressure events, causing accidental intrusion of potentially contaminated water surrounding the pipes. The intrusion volume at potential intrusion nodes is adjusted for nodal pressure and pipe state (age and material) using leakage demand.

- By implementing PDA, the pattern of kitchen tap use was dynamically modified to include the impact of demand availability during PDCs in the analysis. During the PDCs, using a higher critical value of the DSR (5% instead of no demand) for drinking water withdrawals led to a significant reduction in the number of infected people (~65% on Day 1 of 1-h PDCs). This reduction in infection risk if contaminated water is not consumed should be considered to guide preventive notices. It shows that customers should be advised not to drink water when flow at the tap is low (i.e., it takes much longer time to fill a glass).

- In this work, depending on the pathogen concentration in sewage, the number of infected people changed by 235-fold, showing the importance of selecting a representative level of contamination in a system. Using raw sewage as the ingress water is a conservative scenario as water surrounding water mains is likely to be less contaminated than sewage.

- Results show that the number of glasses per day (1, 3, or 10) was less important than the consumption volume (300 mL, 500 mL, or 1 L) for the scenario of 24-h PDCs.

- The duration of PDCs/intrusion is a decisive factor in determining the infection risk, issuing sectorial boil water advisories and other preventive/corrective actions. Spatial and temporal distribution of nodal risks presented in this study can help to determine the boundaries and duration of sectorial BWAs.

- A fast response by the utility is key to reducing the infection risk by limiting the contamination area. For a 1-h intrusion, delaying 5 h the necessary preventive/corrective actions from the start of the intrusion may result in the infection of up to 71 people.

**Supplementary Materials:** The following are available online at http://www.mdpi.com/2073-4441/11/7/1372/s1, Figure S1: Consumption at kitchen tap use, Figure S2: Distribution of population, Figure S3: Probability distribution of the number of infected people during four days of simulation with daily pattern in the hydraulic model, Figure S4: Spatial distribution of pressure.

**Author Contributions:** Conceptualization, F.H., M.P., M.B., M.-C.B., and G.E.; Investigation and formal analysis, F.H. and M.P.; Methodology and software, F.H., and M.B.; writing—original draft, F.H.; writing—review and editing, M.B, M.-C.B, G.E., and M.P.

**Funding:** This research was funded by the NSERC Industrial Chair on Drinking Water at Polytechnique Montréal.

**Acknowledgments:** The participating utility gratefully provided information of the distribution system model. The authors would like to thank Bentley Systems for providing academic access, with unlimited pipes version, to the utility model.

**Conflicts of Interest:** The authors declare no conflict of interest.

## References

1. Kirmeyer, G.J.; Friedman, M.; Martel, K.; Howie, D.; LeChevallier, M.; Abbaszadegan, M.; Karim, M.; Funk, J.; Harbour, J. *Pathogen Intrusion into the Distribution System*; 90835; American Water Works Association Research Foundation, American Water Works Association and United States Environmental Protection Agency: Denver, CO, USA, 2001; p. 254.

2. World Health Organisation (WHO). *Quantitative Microbial Risk Assessment: Application for Water Safety Management*; World Health Organisation: Geneva, Switzerland, 2016; p. 204.

3. Rossman, L.A. *EPANET 2. User's Manual*; EPA 600-R-00-57; National Risk Management Research Laboratory, Office of Research and Development, United States Environmental Protection Agency (USEPA): Cincinnati, OH, USA, September 2000; p. 200.

4. Bashi-Azghadi, S.N.; Afshar, A.; Afshar, M.H. Multi-period response management to contaminated water distribution networks: Dynamic programming versus genetic algorithms. *Eng. Optim.* **2017**, *50*, 415–429. [CrossRef]

5. Rasekh, A.; Brumbelow, K. Drinking water distribution systems contamination management to reduce public health impacts and system service interruptions. *Environ. Model. Softw.* **2014**, *51*, 12–25. [CrossRef]

6. Zafari, M.; Tabesh, M.; Nazif, S. Minimizing the adverse effects of contaminant propagation in water distribution networks considering the pressure-driven analysis method. *J. Water Resour. Plann. Manag.* **2017**, *143*. [CrossRef]

7. Hatam, F.; Besner, M.-C.; Ebacher, G.; Prévost, M. Combining a multispecies water quality and pressure-driven hydraulic analysis to determine areas at risk during sustained pressure-deficient conditions in a distribution system. *J. Water Resour. Plann. Manag.* **2018**, *144*, 04018057. [CrossRef]

8. Lindley, T.R.; Buchberger, S.G. Assessing intrusion susceptibility in distribution systems. *J. Am. Water Works Assoc.* **2002**, *94*, 66–79. [CrossRef]

9. Craun, G.F.; Brunkard, J.M.; Yoder, J.S.; Roberts, V.A.; Carpenter, J.; Wade, T.; Calderon, R.L.; Roberts, J.M.; Beach, M.J.; Roy, S.L. Causes of outbreaks associated with drinking water in the United States from 1971 to 2006. *Clin. Microbiol. Rev.* **2010**, *23*, 507–528. [CrossRef] [PubMed]

10. Viñas, V.; Malm, A.; Pettersson, T.J.R. Overview of microbial risks in water distribution networks and their health consequences: quantification, modelling, trends, and future implications. *Can. J. Civil Eng.* **2019**, *46*, 149–159. [CrossRef]

11. Hamouda, M.A.; Jin, X.; Xu, H.; Chen, F. Quantitative microbial risk assessment and its applications in small water systems: A review. *Sci. Total Environ.* **2018**, *645*, 993–1002. [CrossRef]

12. Besner, M.-C.; Prévost, M.; Regli, S. Assessing the public health risk of microbial intrusion events in distribution systems: conceptual model, available data, and challenges. *Water Res.* **2011**, *45*, 961–979. [CrossRef]

13. Teunis, P.F.M.; Xu, M.; Fleming, K.K.; Yang, J.; Moe, C.L.; LeChevallier, M.W. Enteric virus infection risk from intrusion of sewage into a drinking water distribution network. *Environ. Sci. Technol.* **2010**, *44*, 8561–8566. [CrossRef]

14. Yang, J.; LeChevallier, M.W.; Teunis, P.F.M.; Xu, M. Managing risks from virus intrusion into water distribution systems due to pressure transients. *J. Water Health* **2011**, *9*, 291–305. [CrossRef] [PubMed]

15. LeChevallier, M.W.; Xu, M.; Yang, J.; Teunis, P.; Fleming, K.K. *Managing Distribution System Low Transient Pressures for Water*; Water Research Foundation and American Water Works Service Company, Inc.: Denver, CO, USA, 2011; p. 144.

16. Besner, M.-C.; Messner, M.; Regli, S. Pathogen intrusion in distribution systems: model to assess the potential health risks. Proceedings of 12th Annual Conference on Water Distribution Systems Analysis (WDSA), Tucson, AZ, USA, 12–15 September 2010; pp. 484–493.

17. Davis, M.J.; Janke, R. Development of a probabilistic timing model for the ingestion of tap water. *J. Water Resour. Plann. Manag.* **2009**, *135*, 397–405. [CrossRef]

18. Blokker, M.; Smeets, P.; Medema, G. Quantitative microbial risk assessment of repairs of the drinking water distribution system. *Microb. Risk Anal.* **2018**, *8*, 22–31. [CrossRef]

19. Blokker, M.; Smeets, P.; Medema, G. QMRA in the Drinking Water Distribution System. *Procedia Eng.* **2014**, *89*, 151–159. [CrossRef]

20. Schijven, J.; Forêt, J.M.; Chardon, J.; Teunis, P.; Bouwknegt, M.; Tangena, B. Evaluation of exposure scenarios on intentional microbiological contamination in a drinking water distribution network. *Water Res.* **2016**, *96*, 148–154. [CrossRef] [PubMed]

21. Islam, N.; Rodriguez, M.J.; Farahat, A.; Sadiq, R. Minimizing the impacts of contaminant intrusion in small water distribution networks through booster chlorination optimization. *Stoch. Environ. Res. Risk Assess.* **2017**, *31*, 1759–1775. [CrossRef]

22. Ebacher, G.; Besner, M.-C.; Clément, B.; Prévost, M. Sensitivity analysis of some critical factors affecting simulated intrusion volumes during a low pressure transient event in a full-scale water distribution system. *Water Res.* **2012**, *46*, 4017–4030. [CrossRef] [PubMed]

23. Gibson, J.; Karney, B.; Guo, Y. Predicting health risks from intrusion into drinking water pipes over time. *J. Water Resour. Plann. Manag.* **2019**, *145*, 04019001. [CrossRef]

24. Bentley Systems, Incorporated. *WaterGEMS V8i Users Manual*; Haestad Methods Solution Centre: Watertown, CT, USA, 2014.

25. World Health Organization (WHO). *Risk Assessment of Cryptosporidium in Drinking Water*; WHO/HSE/WSH/09.04; Public Health and Environment, Water, Sanitation, Hygiene and Health; World Health Organization: Geneva, Switzerland, 2009; p. 134.

26. Payment, P.; Plante, R.; Cejka, P. Removal of indicator bacteria, human enteric viruses, Giardia cysts, and Cryptosporidium oocysts at a large wastewater primary treatment facility. *Can. J. Microbiol.* **2001**, *47*, 188–193. [CrossRef]

27. Hatam, F.; Besner, M.-C.; Ebacher, G.; Prévost, M. Improvement of Accidental Intrusion Prediction Due to Sustained Low-Pressure Conditions: Implications for Chlorine and E. coli Monitoring in Distribution Systems. *J. Water Resour. Plann. Manag.*. submitted.

28. Van Zyl, J.E.; Clayton, C.R.I. The effect of pressure on leakage in water distribution systems. *Inst. Civil Eng. Water Manag.* **2007**, *160*, 109–114. [CrossRef]

29. Van Zyl, J.E.; Malde, R. Evaluating the pressure-leakage behaviour of leaks in water pipes. *J. Water Suppl. Resear. Technol. Aqua* **2017**, *66*, 287–299. [CrossRef]

30. Van Zyl, J.E.; Lambert, A.O.; Collins, R. Realistic modeling of leakage and intrusion flows through leak openings in pipes. *J. Hydraul. Eng.* **2017**, *143*. [CrossRef]

31. Besner, M.-C.; Ebacher, G.; Lavoie, J.; Prévost, M. Low and negative pressures in distribution systems: Do they actually result in intrusion? Proceedings of 9th Annual Water Distribution System Analysis Symposium, ASCE-EWRI World Environmental and Water Resources Congress, Tampa, FL, USA, 15–19 May 2007; p. 10.

32. Ebacher, G.; Besner, M.-C.; Prevost, M. Submerged appurtenances and pipelines: An assessment of water levels and contaminant occurrence. *J. Am. Water Works Assoc.* **2013**, *105*, E684–E698. [CrossRef]

33. Besner, M.-C.; Broséus, R.; Lavoie, J.; Di Giovanni, G.; Payment, P.; Prévost, M. Pressure monitoring and characterization of external sources of contamination at the site of the Payment drinking water epidemiological studies. *Environ. Sci. Technol.* **2010**, *44*, 269–277. [CrossRef] [PubMed]

34. Yang, J.; Schneider, O.D.; Jjemba, P.K.; Lechevallier, M.W. Microbial risk modeling for main breaks. *J. Am. Water Works Assoc.* **2015**, *107*, E97–E108. [CrossRef]

35. Van Abel, N.; Blokker, E.J.; Smeets, P.W.; Meschke, J.S.; Medema, G.J. Sensitivity of quantitative microbial risk assessments to assumptions about exposure to multiple consumption events per day. *J. Water Health* **2014**, *12*, 727–735. [CrossRef] [PubMed]

36. Davis, M.J.; Janke, R. Importance of exposure model in estimating impacts when a water distribution system is contaminated. *J. Water Resour. Plann. Manag.* **2008**, *134*, 449–456. [CrossRef]

37. National Research Council of the National Academies. *Drinking Water Distribution Systems: Assessing and Reducing Risks*; The National Academies Press: Washington, DC, USA, 2006; p. 404.

*Article*

# Efficacy of Flushing and Chlorination in Removing Microorganisms from a Pilot Drinking Water Distribution System

**Nikki van Bel [1],*, Luc M. Hornstra [1], Anita van der Veen [1] and Gertjan Medema [1,2]**

[1]  KWR Watercycle Research Institute, P.O. Box 1072, 3430 BB Nieuwegein, The Netherlands;
    Luc.Hornstra@kwrwater.nl (L.M.H.); Anita.van.der.Veen@kwrwater.nl (A.v.d.V.);
    Gertjan.Medema@kwrwater.nl (G.M.)
[2]  Sanitary Engineering, Department of Water Management, Faculty of Civil Engineering and Geosciences,
    Delft University of Technology, P.O. Box 5048, 2600 GA Delft, The Netherlands
*  Correspondence: Nikki.van.Bel@kwrwater.nl; Tel.: +31-30-606-9516

Received: 22 March 2019; Accepted: 24 April 2019; Published: 29 April 2019

**Abstract:** To ensure delivery of microbiologically safe drinking water, the physical integrity of the distribution system is an important control measure. During repair works or an incident the drinking water pipe is open and microbiologically contaminated water or soil may enter. Before taking the pipe back into service it must be cleaned. The efficacy of flushing and shock chlorination was tested using a model pipe-loop system with a natural or cultured biofilm to which a microbial contamination (*Escherichia coli*, *Clostridium perfringens* spores and phiX174) was added. On average, flushing removed 1.5–2.7 log microorganisms from the water, but not the biofilm. In addition, sand added to the system was not completely removed. Flushing velocity (0.3 or 1.5 m/s) did not affect the efficacy. Shock chlorination (10 mg/L, 1–24 h) was very effective against *E. coli* and phiX174, but *C. perfringens* spores were partly resistant. Chlorination was slightly more effective in pipes with a natural compared to a cultured biofilm. Flushing alone is thus not sufficient after high risk repair works or incidents, and shock chlorination should be considered to remove microorganisms to ensure microbiologically safe drinking water. Prevention via hygienic working procedures, localizing and isolating the contamination source and issuing boil water advisories remain important, especially during confirmed contamination events.

**Keywords:** chlorination; flushing; drinking water distribution system; water quality; contamination; cleaning

---

## 1. Introduction

Due to extensive purification of ground water or surface water, the level of pathogenic microorganisms and indicator organisms are absent in detectable levels in treated drinking water. To ensure that microbial safety of drinking water is maintained during distribution, the two most important control measures are the physical integrity of the distribution system and a continuously high water-pressure. These measures combined prevent the intrusion of contaminated water carrying pathogenic microorganisms in the distribution system. If these two requirements are not met, a chlorine residual in the distribution system may protect the drinking water to some extent to intrusion of microorganisms. The drinking water in the Netherlands is distributed without a chlorine disinfection residual, which requires extra vigilance during repair works or incidents. If small cracks are present and the water pressure is reduced, for example due to a pipe break or pressure transients, ground water may leak into the distribution system [1]. During repair works or incidents contamination may occur as well, despite the hygienic procedures that are in place. Sewage and drinking water pipes are often present in the underground and are close to each other. For that reason the soil and ground water next to drinking water pipes contain high numbers of microorganisms (median values): fecal coliforms (2–59 × $10^1$ MPN/100 mL, 2 × $10^1$–1 × $10^2$ MPN/100 g), *Clostridium perfringens* (5 × $10^1$–1 × $10^3$ cfu/100 mL; 1 × $10^1$–1 × $10^3$ cfu/100 g), *Bacillus subtilis* (1.3 × $10^6$ cfu/100 mL; 1.3 × $10^8$ cfu/100 g) and coliphages (1 × $10^4$ pfu/100 mL; absent in soil) [1,2]. Intrusion of ground water into the distribution system may thus lead to contamination of the drinking water and distribution system with enteric pathogens. Even intrusion of a small volume into the distribution system may be of concern as several enteric pathogens are highly infectious [3,4].

Several studies have shown that incidents or repair works of the drinking water distribution system are associated with an elevated risk of gastrointestinal diseases. A cohort study in Norway among households downstream of a main break or maintenance works were compared to unexposed households and showed that exposed households reported 1.58 times more gastrointestinal illnesses [5]. A cohort study in Sweden showed that households were 2.0 times as likely to report vomiting complaints and 1.9 times as likely to report acute gastrointestinal illness after a pipe break or works on the distribution system [6]. Risk factors were identified and included the presence of sewage pipelines at the same level as drinking water pipelines in the trench. Flushing was also associated with an elevated risk, leading to the conclusion that current safety measures might not be sufficient in eliminating the risk of gastrointestinal diseases. Households in Canada consuming tap water reported 19% to 34% more gastrointestinal illnesses compared to households receiving water bottled directly at the production plant or receiving tap water treated with reverse osmosis or receiving spring water [7,8]. In addition, in a Swedish study routinely available data on the incidence of Campylobacter was used and shown to be positively associated with the average water-pipe length per person suggesting contamination of the water during distribution [9]. Both studies suggest that during distribution the water was contaminated with pathogens, although the distribution systems did meet the microbiological standards (monitoring of indicator bacteria). In contrast, no proof was found for contamination during distribution in two blinded studies: no difference was found for the risk of waterborne gastroenteritis between households in Melbourne [10] or Iowa [11] using a real or sham water treatment unit installed in the kitchen. In a systematic review, a significant association between gastrointestinal disease and tap water versus point of use treated tap water was found for non-blinded studies. However these differences disappeared when only blinded studies were included [12].

To estimate the potential health impacts of contamination of the drinking water distribution system, the QMRA (Quantitative Microbial Risk Assessment) approach was applied [13–16]. A QMRA of negative pressure transients showed that the health risk was mainly affected by the duration of a negative pressure event [14,15] and by the number of nodes drawing negative pressure. The concentration of the contaminant, in both studies Norovirus was used, was not critical, probably due to the high infectivity of the virus [3]. In a QMRA of the health impact of contamination during repair works, the concentration of the contaminant, the time of day the valves are opened after repairs, and the time of consumption were the most important parameters [17,18]. Additionally, the type of pathogen and its specific dose–response relationship highly impacts the resulting infection risk [18]. For example, ingestion of roughly 10–100× higher numbers of bacteria compared to protozoa, or 100–1000× higher compared to viruses, are required to become infected [19–21]. Together these studies suggest a (potential) role for the loss of physical integrity, pipe breaks and repair works of the drinking water distribution system in the transmission of gastrointestinal diseases.

Flushing and, in case of a persistent microbial contamination in the distribution system, shock chlorination are often used to clean the distribution system after maintenance work or pipe breaks. However, to our knowledge, only limited research has been performed to test the efficacy of flushing and shock chlorination on microorganisms in water and pipe wall biofilm. Flushing experiments in a pilot distribution system used sand particles with a diameter of 0.25–4 mm and showed that a threshold velocity of 0.8–0.9 m/s was required to achieve efficient removal [22]. In the same study the results were extrapolated to the removal of microorganisms. Chlorination was tested in a batch reactor, in which the microorganisms were present in the water but not in a biofilm. Flushing (0.8 m/s) does not remove *Bacillus subtilis* from the biofilm of a cement-lined ductile iron pipe, whereas shock chlorination (200 mg/L, 2 h) in the same pilot distribution system led to a modest 1.2–1.4 log removal from the biofilm [23]. However, removal from the water was not tested.

After cleaning the distribution system using flushing or chlorination, drinking water companies want to put the distribution system back into service as quickly as possible. Current practice is that the cleaned area can be put back into service after the monitoring of the water for the fecal indicators *E. coli* and enterococci shows that these are absent. To ensure the safety of the consumers and to be able to restore normal water distribution, it is important that monitoring can be performed quickly after the cleaning regime is finished and that monitoring is optimized in such a way that the probability of detecting fecal indicators (if present) is as high as possible. Currently, a sample is taken from a convenient tap in the vicinity of the cleaned network, but information about the impact of time and place of sampling on the probability of picking up residual fecal indicators is lacking.

In this paper we used a model pipe-loop system, in which a biofilm was cultured, to study (i) the influence of the waiting time between flushing of the system and sampling the water and (ii) the effect of the distance between sampling point and point of contamination. This knowledge is used to underpin the sampling strategy after flushing in which the waiting time should be as short as possible without decreasing the probability of detecting the fecal indicators. We extend the knowledge on the efficacy of flushing and shock chlorination in removal of a microbial contamination from drinking water pipes. Both flushing and shock chlorination were tested in a model pipe-loop system containing pipes with a cultured biofilm or pipes with an old and natural biofilm that were excavated from the drinking water distribution system. Flushing was tested using several conditions (flushing velocity and flushed water volume) and chlorination was tested for several time periods. Following this, we describe the application of these results to real-life situations.

## 2. Materials and Methods

Experiments were carried out with Escherichia coli (*E. coli* WR1, NCTC 13167), *Enterococcus faecium* (*E. faecium* WR63, NCTC 13169), *Clostridium perfringens* (*C. perfringens* D10, NCTC 13170) spores, somatic coliphage phiX174 (ATCC 13706-B1) and bacteriophage MS2 (ATCC 15597-B1). *E. coli* is the most commonly studied microorganism and also the one most often used as a bacterial indicator of fecal pollution in drinking water. Bacteriophages are used in this study as an alternative for enteric viruses, due to their match in morphology and biological properties [24]. The efficacy of chlorine disinfection of the distribution system depends on several factors including the chlorine concentration, contact time and the type of microorganism. Therefore several types of microorganisms were selected representing bacteria (*E. coli*), bacterial spores (*C. perfringens* spores) and viruses (the bacteriophage phiX174), and also representing a spectrum of chlorine-sensitivity, from sensitive (*E. coli*) to insensitive (spores of *C. perfringens*). For analysis of the optimal sampling strategy, the fecal indicator bacteria *E. coli* and enterococci were used. These two indicators are routinely monitored and monitoring is compulsory after works in the distribution system. A water sample of 100 mL should be negative for both bacteria.

*E. coli* and *Enterococcus faecium* used to test the optimal sampling strategy (experiments A1–A4, Table 1), were grown in mineral medium supplied with glucose and potassium nitrate, respectively glucose and brain heart infusion broth. Growth at 22 °C was monitored by streak-plating on Lab Lemco Agar (LLA) plates. When the maximum colony count was reached the bacterial suspensions were stored at 4 °C until use. One suspension of each bacterium was prepared and used in all experiments. Shortly before each experiment the colony counts were determined on a specific medium: Laurylsulphate agar (LSA) for *E. coli* and Slanetz and Bartley agar (S&B) for *Enterococcus faecium*. The *E. coli* colony counts were determined according to NEN-EN-ISO 9308-1 using membrane filtration (0.45 μm pore size), or streak-plating of the sample on LSA-agar plates. Agar plates were incubated for 5 h at 25 °C followed by 14 h at 36 °C. The *Enterococcus faecium* colony counts were determined on S&B agar for 48 h at 36 °C, according to NEN-EN-ISO 7899-2. MS2 F-specific bacteriophages were purchased via GAP

EnviroMicrobial Services. The number of MS2 was determined according to the double-agar layer method in NEN-ISO 10705-1. In short, decimal dilutions of the sample containing MS2 was mixed with the host bacterium *Salmonella typhimurium* WG49 in log-phase and semi-solid Tryptone-Yeast Extract-Glucose agar. The mixture was immediately spread on an agar plate and allowed to solidify. Incubation was performed at 36 °C for 24 h. The heterotrophic plate count (HPC) was determined according to NEN-EN-ISO 6222. The sample is mixed with dissolved plate count agar solution and poured into a petri dish. The sample is incubated at 22 °C for 68 h.

**Table 1.** Flushing conditions to determine the optimal sampling strategy.

|  | Contamination | *E. coli* Total cfu | *Enterococci* Total cfu | MS2 Total pfu | Flushing |
|---|---|---|---|---|---|
| A1 | Water | $1.2 \times 10^{10}$ | $8.1 \times 10^8$ | $1.3 \times 10^{11}$ | 1.27 m/s, 2.5 vol |
| A2 | Water | $5.3 \times 10^9$ | $4.0 \times 10^8$ | $9.7 \times 10^{10}$ | 1.27 m/s, 2.5 vol |
| A3 | Sand and Water | $1.1 \times 10^{10}$ | $2.6 \times 10^8$ | $1.7 \times 10^{11}$ | 1.27 m/s, 2.5 vol |
| A4 | Water | $2.6 \times 10^{10}$ | $7.4 \times 10^8$ | $1.0 \times 10^{11}$ | 0.45 m/s, 0.9 vol |

For the flushing and chlorination experiments (experiments B1–B5 and C1–C3, Tables 2 and 3), *E. coli* bacteria were grown in Lab Lemco Broth (LLB) for 72 h at 36 °C. Growth medium was removed by washing the bacteria three times in sterile tap water. The number of *E. coli* bacteria in the suspension was determined on LSA agar plates as described above. For each experiment the bacteria were freshly prepared. *C. perfringens* D10 was cultured on Perfringens agar base (PAB), after which the colonies were aseptically transferred to sterile tap water. To induce sporulation of the bacteria, the bacteria were incubated for two weeks at 36 °C after which the spores were stored at 4 °C. One spore suspension was prepared and used for all experiments. A few days before each experiment the number of spores in the suspension was determined according to NEN-ISO 6461. In short, prior to enumeration the sample was heated at 60 °C for 30 min to kill vegetative bacteria. The colony count was determined using membrane filtration or streak-plating on PAB agar plates at 36 °C for 24 to 48 h. phiX174 somatic coliphages were grown by infection of an *E. coli* WG5 culture with phiX174 for 5 h at 36 °C. The growth medium was removed by centrifugation and ultrafiltration. One suspension of phiX174 coliphages was prepared and used for all experiments. A few days before each experiment the number of phiX174 in the suspension was determined according to NEN-ISO 10705-2. The sample was mixed with *E. coli* WG5 in log-phase and semi-solid Modified Scholtens Agar, spread immediately on an agar plate and allowed to solidify. Incubation was performed at 36 °C for 24 h.

**Table 2.** Conditions to determine the efficacy of flushing and chlorination on the removal microorganisms from the water and biofilm.

|  | *E. coli* Total cfu | cfu/mL | *Clostridium D10* Total cfu | cfu/mL | phiX174 Total pfu | pfu/mL | Flushing | Free Chlorine |
|---|---|---|---|---|---|---|---|---|
| B1 | $7.5 \times 10^{10}$ | $9.4 \times 10^5$ | $2.6 \times 10^8$ | $3.3 \times 10^3$ | - | - | 1.5 m/s, 3 vol | 10 mg/L, 0–24 h |
| B2 | $6.3 \times 10^{10}$ | $5.7 \times 10^5$ | $8.4 \times 10^8$ | $7.6 \times 10^3$ | $1.0 \times 10^{11}$ | $9.1 \times 10^5$ | 1.5 m/s, 3 vol | 10.5 mg/L, 0–24 h |
| B3 | $7.9 \times 10^{10}$ | $7.2 \times 10^5$ | $2.0 \times 10^9$ | $1.8 \times 10^4$ | $5.4 \times 10^{10}$ | $5.0 \times 10^5$ | 1.0 m/s, 3 vol | – |
| B4 | $7.5 \times 10^{11}$ | $6.8 \times 10^6$ | $3.4 \times 10^8$ | $3.1 \times 10^3$ | $9.4 \times 10^{10}$ | $8.5 \times 10^5$ | 0.3 m/s, 3–6–10–15 vol | 10 mg/L, 0–24 h |
| B5 | $5.2 \times 10^8$ | $5.7 \times 10^{10}$ | $1.5 \times 10^7$ | $1.7 \times 10^9$ | $4.0 \times 10^8$ | $4.4 \times 10^{10}$ | 0.3 m/s, 3–6–10–15 vol; 1.5 m/s, 3–6–10–15 vol | - |

**Table 3.** Experimental conditions to determine the efficacy of flushing and chlorination on the removal of microorganisms from the water and biofilm from real-world pipe segments with a natural biofilm.

| | *E. coli* | | *C. perfringens* | | phiX174 | | Flushing | Free Chlorine |
|---|---|---|---|---|---|---|---|---|
| | Total cfu | cfu/mL | Total cfu | cfu/mL | Total pfu | pfu/mL | | |
| C1 | $5.0 \times 10^{10}$ | $4.5 \times 10^5$ | $1.0 \times 10^9$ | $9.2 \times 10^9$ | $6.7 \times 10^{10}$ | $6.1 \times 10^5$ | 1.5 m/s, 3 vol | 10 mg/L, 0–24 h |
| C2 | $7.5 \times 10^{10}$ | $5.8 \times 10^5$ | $3.2 \times 10^9$ | $2.9 \times 10^4$ | $4.0 \times 10^{10}$ | $3.7 \times 10^5$ | 1.5 m/s, 3 vol | 10 mg/L, 0–24 h |
| C3 | $1.0 \times 10^{11}$ | $9.5 \times 10^5$ | $3.0 \times 10^9$ | $2.7 \times 10^4$ | $4.7 \times 10^{10}$ | $4.3 \times 10^5$ | 0.3 m/s, 3–6–10–15 vol | 10 mg/L, 0–24 h |

To determine the adenosine triphosphate (ATP) concentration in the samples luciferin and luciferase were added. In the presence of ATP luciferin is degraded by luciferase during this process light is produced and measured in a luminometer. For measuring the iron concentration the water sample was treated with nitric acid according to NEN-EN-ISO 15587-2:2002. The concentration of the released iron determined using inductively coupled plasma mass spectrometer (ICP-MS).

Analyses of the biofilm were performed by swabbing roughly 7 cm of pipe wall biofilm (all sides), 1–2 cm pipe wall biofilm at the start of the pipe segment was not swabbed. For swabbing multiple sterile cotton swabs were used. Of each pipe the swabbed surface area was calculated. The swabs were pooled in 40 mL sterile tap water and the biomass was released from the swab by low-energy sonication using a Branson Sonifier ultrasonic cell disruptor for 2 min at 40-kHz and 90-Watt power output (equivalent to 45% amplitude). During sonification the mixture was kept on ice. In the resulting water sample the required parameters were determined.

Die-off kinetics of all microorganisms were determined in tap water to which a concentration of microorganisms was added in similar concentrations as to the pipe-loop system. The water was incubated for 24–72 h at 22 °C, the number of microorganisms was determined several times during this period.

In some of the experiments an artificial biofilm was cultured in the pipe-loop system prior to the experiment. To achieve a reproducible biofilm acetate (NaCH$_3$COO, 10 µg C/L), nitrate (KNO$_3$, 2 µg N/L) and phosphate (KH$_2$PO$_4$, 0.1 µg P/L) were added to the water. After continuous circulation of the water at 0.1 m/s in the pilot system for two weeks the biofilm and water were analyzed for ATP and HPC. The system was drained and flushed on low speed to remove remaining nutrients before starting the experiment.

Three slightly different model pipe-loop systems were built (Figure 1). All systems were 20 m long and consisted of PVC-pipes with an outer diameter of 63 mm and inner diameter of 55 mm. At several points along the 20 m-system taps for water sampling were placed. A flow meter was placed immediately after the flushing pump. After each experiment the biofilm was removed from the complete system by flushing with a SDS solution, tap water, citric acid solution and tap water. In addition, the long stretches of pipe were replaced with new pipes. Depending on the specific research questions, small alterations were made to the pipe-loop system, as described below.

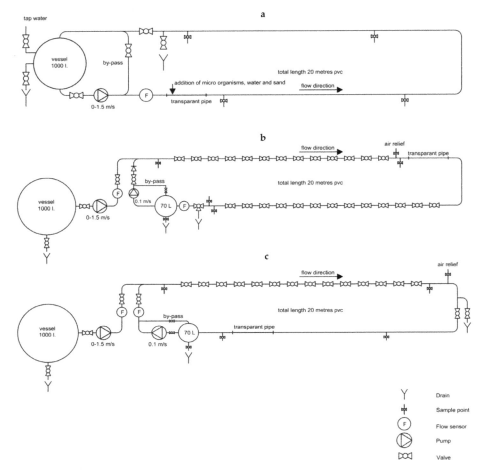

**Figure 1.** Schematic overview of the pipe-loop distribution systems that were used for experiments A1–A4 (**a**), B1–B3 (**b**) and B4, B5, C1–C3 (**c**). In systems B and C pipe segments were placed between valves that were replaced with new segments to sample the biofilm.

For determining the optimal sampling strategy (experiments A1–A4), the system in Figure 1a was used. Sampling taps were placed after 1, 5, 10 and 20 m. For all experiments a biofilm was cultured. Next, the system was drained and artificially contaminated water or sand was added to the system immediately after the flushing pump and flow meter. The contaminated water or sand was left in the drained pipe for one hour to allow the microorganisms to attach to the biofilm. Flushing was performed according to Table 1. This procedure was chosen to mimic repair works on the distribution system as much as possible (i.e., the opening of a drained pipe and contaminated water or sand that may enter and reside for some time in the pipe before flushing is started). After flushing the system was closed, without any circulation or flow of the water to mimic the withholding time of the water in the pipe before the sample is taken, and water samples were taken after 1, 2, 6 and 24 h at the four sampling taps. At the end of the experiment the system was drained and the biofilm was sampled at 1, 10, and 20 m to determine the amount of microorganisms present in the biofilm. The efficacy of the several flushing regimes on this pipe-loop system was compared by calculating a mass balance. It was assumed that before flushing all microorganisms were present in the water and were absent from the biofilm. Numbers found in the water and biofilm samples 24 h after flushing were extrapolated to the entire water phase and pipe wall. Using these numbers the log removal of the microorganisms was calculated.

To determine to efficacy of flushing and chlorination on removal of microorganisms from water and biofilm (experiments B1–B5), the system in Figure 1b was used. A large number of removable pipe segments were included with ball valves on both sides. Individual segments could be taken out for analyses and replaced by a new segment, without loss of water or the need to drain the system. A small circulation reservoir, circulation pump, and flow meter were included in the system. During the entire experiment the water was circulated with 0.1 m/s (the average flow in the Dutch distribution system) using the circulation pump. For all experiments a biofilm was cultured. Microorganisms were added to the water and circulated for 24 h before the system was flushed according to Table 2. Only the pipes between the flushing pump and the drain to the sewage system were flushed. The parts necessary for circulation (pump, reservoir, flow meter) were blocked by closed valves. After flushing, samples were taken and the water in the entire system was circulated for 30 min to ensure mixing of microorganisms throughout the water phase. Shock chlorination was started by addition of concentrated sodium hypochlorite to the circulation reservoir up to a concentration of 10 mg/L in the water of the entire system. The free chlorine concentration was measured after 20 min, 1–3–6–24 h (Hach, LCK310). During the flushing and chlorination procedures the water and biofilm were sampled at different time points, depending on the experiment: after 3–6–10–15 flushing volumes (the volume of the pipe segment that is flushed, i.e., 3 flushing volumes implies that the volume of the pipe segment was replaced three times) and after 1, 6 and 24 h contact time with chlorine.

To test the efficacy of flushing and chlorination on real-world pipes (PVC) with a natural biofilm (experiments C1–C3), the system of Figure 1b was adapted in such a way that pipes excavated from a distribution system could be incorporated in the pipe-loop system. The pipes were derived from different Dutch drinking water companies and were 35, 51, and 27 years old. Five meters of pipe, in segments of 50 cm, were excavated under hygienic procedures. The segments were closed with a cap and packaged in plastic bags, transported at 4 °C to the laboratory and incorporated within 6 h in the pilot distribution system. Before incorporation the sawdust was gently removed and the segments were incorporated using the same orientation (flow direction and up and bottom side of the pipe) as in the distribution system. Only the incorporated part was flushed (Figure 1c). No biofilm was grown in the remaining 15 m pipe that were not flushed, to ensure that only the effect of a natural biofilm was monitored. After incorporation of the real-world pipe segments in the system, the system was circulated for 4 days with location-specific drinking water. After 4 days the water was refreshed, microorganisms were added and circulated for 24 h at 0.1 m/s. Flushing and chlorination were performed as described above (Table 3).

## 3. Results and Discussion

### 3.1. Growth or Decay of Microorganisms in Pipe-Loop System

During the experiments with the pipe-loop systems, the microorganisms are present in the pipe-loop system for 24 to 48 h which may induce growth or die-off of the microorganisms. To ensure that an increase or decrease in the number of microorganisms can be attributed to flushing, chlorination, attachment to or release from the biofilm, and is not due to decay (or growth) within this time frame, die-off kinetics of all microorganisms in water were determined (Figure 2). MS2 and enterococci remained stable for at least 24 h, the number of *E. coli*, *C. perfringens* and phiX174 also remained stable for 72 h. Therefore, any changes in the number of microorganisms in the pipe-loop system experiments can be attributed to flushing, chlorination, or attachment to or release from the biofilm.

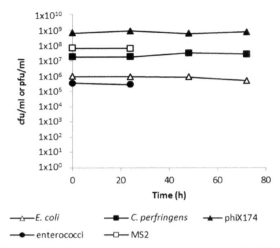

**Figure 2.** Die-off curves of micro-organisms in water for 24–72 h.

## 3.2. Biofilm Characteristics

During all experiments, a biofilm was present in the pipe-loop system, either a biofilm was cultured or real-world pipe segments with a natural biofilm were incorporated in the system. Just before the start of each experiment, the water in the pipe-loop system and the biofilm were analyzed for HPC, ATP and iron. Previous studies mention different HPC numbers for a representative, living biofilm: $10^4$–$10^5$ cfu/mL in water and $10^4$–$10^5$ cfu/cm$^2$ for a cultured biofilm (USEPA, 2008) and $10^1$–$10^6$ cfu/cm$^2$ for established biofilms [25,26]. The HPC count in the water of the pipe-loop systems with a cultured biofilm falls within this range ($5.9 \times 10^4$ cfu/mL), the biofilm yields slightly higher HPC numbers ($3.1 \times 10^5$ cfu/cm$^2$; Figure 3A). ATP counts were high ($2.8 \times 10^4$ pg/mL and $2.9 \times 10^4$ pg/cm$^2$), indicating a large active biomass population and a very low amount of iron was present (0.23 μg/cm$^2$). The low iron concentration shows that within two weeks a microbiological representative and living biofilm can be cultured, but accumulation of iron in the biofilm is very limited. The excavated pipes from the distribution system with a natural biofilm showed lower HPC counts and ATP concentrations than the cultured biofilms ('Entry' in Figure 3B): 16.7 cfu HPC/mL and 3.97 pg ATP/mL in the water and 40.2 cfu HPC/mL and 227 pg ATP/cm$^2$ in the biofilms. The iron concentration was much higher (7.5 μg/cm$^2$) and clearly visible by eye. After four days of circulation with drinking water, the HPC number had increased with 3.2–4.8 logs and the ATP concentration with 0.7–1.6 log. The iron and TOC levels remained stable at 5.4–7.0 μg/cm$^2$ respectively 0.01 mg C/cm$^2$.

## 3.3. Flushing with 1.27 m/s

To test the optimal sampling strategy after flushing several contamination scenarios were simulated in the pipe-loop system (Figure 1a, Table 2). Flushing according to the guidelines (1.27 m/s, 2.5 volumes) removed 2.7–3.4 log *E. coli*, 3.5–4.2 log enterococci and 3.5–3.9 log MS2 from the water (Figure 4A, experiments A1 and A2). Microorganisms in sand and water were removed to a comparable extent (A3). Flushing with suboptimal velocity (0.45 m/s) and replacing the water with only 0.9 volumes is much less efficient (experiment A4). Until now it was generally assumed that flushing a distribution system with a clean water front removes all contaminants such as sediment and microorganisms. Although a large part of the microorganisms was removed, the results show that the pipe-loop system was not entirely clean and microorganisms remained in both water and biofilm, despite the high flushing speed and volumes that were used.

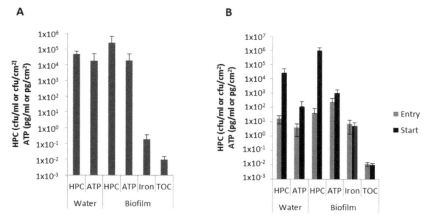

**Figure 3.** Average (with standard deviation (SD)) of heterotrophic plate count (HPC) number, adenosine triphosphate (ATP) and iron concentration of the cultured biofilm after 14 days ((**A**), n = 2–5) and, including TOC levels, of natural biofilms upon arrival at the laboratory (entry) and immediately before start of the experiment ((**B**), n = 3).

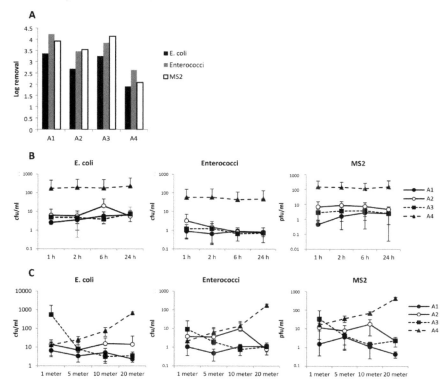

**Figure 4.** (**A**). Flushing efficacy (A1–A3: 1.27 m/s; 2.5 volumes; A4: 0.45 m/s, 0.9 volumes) of the individual experiments. (**B**) Role of waiting time after flushing (**C**) and distance from the contamination point on the number of microorganisms in the water. Shown is the average, with SD, of the results of four distances per time point (**B**) or the average of the results of four time points per distance point (**C**).

## 3.4. Role of Waiting Time between Flushing and Sampling

After flushing, the pipe-loop system was closed and water remained stagnant for 24 h during which samples were taken. The number of microorganisms in the water changed very little over time (Figure 4B). Two exceptions showed a decreasing (enterococci in A2) or increasing (MS2 in A1) pattern. This result was only seen once and therefore suggests that this may be variation caused by the large setup of the experiments or might be explained by natural variation. Differences in waiting time between flushing and sampling thus does not appear to affect the number of microorganisms present in the water. The general lack of changes in the number of microorganisms suggests that exchange between water and biofilm is not significant: the number of microorganisms that attaches to or releases from the biofilm is too small to detect in this setup. An alternative explanation could be that the microorganisms in biofilm and water are in equilibrium, with the same number of microorganisms attaching to and releasing from the biofilm. The results indicate that the probability of detecting fecal indicators one hour after flushing is similar to later time points. Hence, in a real situation in which repair works have been performed, followed by flushing, a waiting time longer than one hour before sampling is not necessary, does not change the chance of detecting a microorganism and unnecessarily leads to a longer period during which that particular section of the drinking water distribution system cannot be released for consumption.

After flushing, the water was sampled at different distances along the pipe-loop system to test whether the distance between the point of contamination and sampling point affects the number of microorganisms that were found. When microorganisms were dosed in water and flushing was performed according to the guidelines, the number of microorganisms in water varied little over the distance (Figure 4C; experiment A1 and A2). Exceptions are *Enterococcus faecium* and MS2 in experiment A2: at 20 m lower numbers were found than at the 10-m sampling point. Dosage of microorganisms in sand and water in experiment A3 leads to high numbers close to the contamination point and the number of microorganisms declined with an increasing distance. Especially at the first sampling point at 1 m high numbers of microorganisms were found, and these numbers declined rapidly with increasing distance. After flushing, the sand was not completely removed and was particularly present in the first few meters after the contamination point, and was most likely the cause of the larger number of microorganisms at 1- and 5-m sampling points. Slower flushing with less volumes yielded the reverse picture, the number of microorganisms increased with the distance (experiment A4). At 1 m the number was comparable to the other experiments, but at 20 m the number was about two logs higher. The water in the pipe-loop system was not completely refreshed as flushing was performed with only 0.9 volumes. After flushing the bulk of the contamination has arrived at around the 20-m point where the sample was taken. Whether the less efficient removal is caused by the lower flushing velocity or by lower flushing volume cannot be determined from these experiments. However, the comparable low number of microorganisms at 1 m between experiments A1, A2 and A4 suggests that the flushing speed is less important.

## 3.5. Attachment of Microorganisms to the Biofilm

In the previous experiments only the flushing efficacy in removing microorganisms from the water was tested. However, in all pipes of the drinking water distribution system a biofilm is present. It may be possible that characteristics of the biofilm influence the attachment of microorganisms and thus the efficacy of flushing and chlorination. Therefore, not only a biofilm was cultured in the pipe-loop system, but also segments of real-world pipes with a natural biofilm were taken from the distribution system and incorporated in the pipe-loop system. Binding of microorganisms to the biofilm, flushing and chlorination were tested in pipe-loop systems with a cultured biofilm (experiments B1–B3, Figure 1b and Table 2) or a natural biofilm (experiments C1–C3, Figure 1c and Table 3).

After addition of the microorganisms to the water in the pipe-loop system, the water was circulated for 24 h to allow the microorganisms to attach to the biofilm (Table 4). The binding efficacy differed for each microorganism, but was comparable between the cultured and natural biofilm: *C. perfringens* spores bound most efficiently to the biofilm (42.3% and 36.4%, respectively), followed by *E. coli* (6.7% and 3.3%) and phiX174 (0.6% and 0.7%). These binding efficiencies are comparable to a previous study using small reactors with glass pearls on which a biofilm was cultured using similar methods as this study. Approximately 10% of the *E. coli* bacteria was bound to the biofilm [27]. However, using young, natural, 7–10 months old, drinking water biofilms in a flow chamber only 0.03% of *E. coli* was attached [28]. Another study showed a 1 log/cm$^2$ binding of phiX174 (after dosing 5.6 log) to a natural, 7-month old, biofilm [29], compared to 3–4 log/cm$^2$ in this study. Additionally, in a pilot distribution system spores of the bacterium *Bacillus subtilis* attached to a cultured biofilm with 3–4 log/cm$^2$ [23]. Some reasons for the large differences in binding characteristics between the studies might be the environment, the composition of the drinking water and the pipe material that affect biofilm formation. However, within our study we did not find differences between a cultured biofilm and a natural biofilm and when drinking water from different locations was used. A clear explanation for the differences is lacking, however, in all studies the microorganisms bind to the biofilm. These attached microorganisms may negatively affect the efficacy of flushing procedures and the chlorine disinfection.

**Table 4.** Binding of microorganisms to cultured and natural biofilm.

| | | E. coli | | C. perfringens | | phiX174 | |
|---|---|---|---|---|---|---|---|
| | | Water | Biofilm | Water | Biofilm | Water | Biofilm |
| | B1 | 99.7 | 0.3 | 27.0 | 73.0 | - | - |
| | B2 | 87.5 | 12.5 | 72.6 | 27.4 | 99.3 | 0.7 |
| Cultured | B3 | 86.3 | 13.7 | 76.2 | 23.8 | 99.7 | 0.3 |
| biofilm | B4 | 93.7 | 6.3 | 28.2 | 71.8 | 98.7 | 1.3 |
| | B5 | 99.2 | 0.8 | 84.4 | 15.6 | 99.9 | 0.1 |
| | Average ± SD | 93.3 ± 6.3 | 6.7 ± 6.3 | 57.7 ± 27.8 | 42.3 ± 27.8 | 99.4 ± 0.5 | 0.6 ± 0.5 |
| | C1 | 94.4 | 5.6 | 45.0 | 55.0 | 99.3 | 0.7 |
| Natural | C2 | 99.1 | 0.9 | 91.8 | 8.2 | 99.8 | 0.2 |
| biofilm | C3 | 96.5 | 3.5 | 54.0 | 46.0 | 98.8 | 1.2 |
| | Average ± SD | 96.7 ± 2.4 | 3.3 ± 2.4 | 63.6 ± 24.9 | 36.4 ± 24.9 | 99.3 ± 0.5 | 0.7 ± 0.5 |

Note: SD: Standard deviation.

### 3.6. Normal Flushing with 1–1.5 m/s

After 24 h circulation, flushing (1.0–1.5 m/s and 3 volumes) removed 1.2–3.1 log *E. coli*, 0.5–3.0 log *C. perfringens* spores and 2.7–3.3 log phiX174 from the water in the presence of a cultured biofilm (Figure 5A). Removal of phiX174 from water seemed to be most efficient, although the differences were small. The log removal in experiment B1 was markedly lower compared to the other experiments, the reason for this is not known. Removal of microorganisms attached to a cultured biofilm by flushing was limited: 0.3–1.3 log *E. coli*, 0–0.9 log *C. perfringens* and 0.9–1.4 log phiX174 were removed. Removal of microorganisms from water and biofilm when pipe segments with a natural, old biofilm were used, was comparable to the experiments with a cultured biofilm (Figure 5B). Removal from the water phase was slightly lower in this case, and relatively constant, despite the differences in biofilm site and age and water quality (1.6–2.6 log *E. coli*; 2.0–2.2 log *C. perfringens*; 2.6–2.7 log phiX174). Removal of microorganisms from the natural biofilm was also slightly lower: 0.2–1.4 log *E. coli*; 0.2–0.4 log *C. perfringens*; −0.2–0.5 log phiX174. Extended flushing with up to 15 volumes only yielded an extra 0.4–0.7 log removal from the water (Figure 6C). Removal in the first three volumes was largest, after which flushing became less effective.

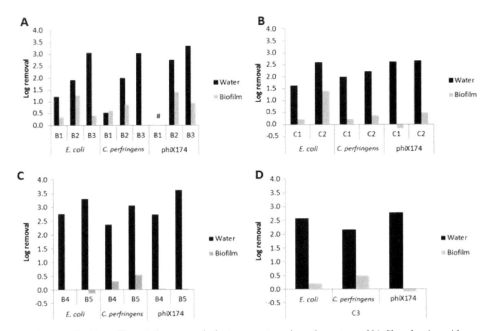

**Figure 5.** Flushing efficacy in log removal of microorganisms from the water and biofilm of a pipe with a cultured biofilm (**A,C**) or a natural biofilm (**B,D**). In (**A**) (experiments B1–B3) and (**B**) (experiments C1–C2) the pipe-loop system was flushed with 3 volumes at 1.5 m/s. In (**C**) (experiment B4–B5) and (**D**) (experiment C3) the pipe-loop system was flushed with 3 (water) or 15 volumes (biofilm) at 0.3 m/s. # phiX174 not added to pipe-loop system and not determined.

*3.7. Slow Flushing with 0.3 m/s*

Although flushing of the drinking water distribution system should normally be performed at a velocity of 1.5 m/s and 3 volumes, this is not always possible due to practical reasons. As mentioned in the AWWA standard C651-14, an alternative can be to flush the distribution system with a lower velocity [30]. In addition, flushing can be extended so more pipe volumes are replaced. However, it is unknown whether this is as efficient as flushing with 1.5 m/s. Flushing with low speed (0.3 m/s) was tested in the pipe-loop system with a cultured (experiments B4–B5) or natural biofilm (experiment C3). Flushing with three volumes at 0.3 m/s removed 2.4–2.8 log microorganisms from the water of a pipe-loop system with a cultured biofilm (Figure 6A). Microorganisms were not removed from the cultured biofilm (0–0.3 log removal). Flushing pipes at low speed with three volumes and a natural biofilm yielded comparable results: 2.2–2.8 log removal of microorganisms from water and −0.1–0.5 log removal from the biofilm (Figure 6B). Extended flushing with more volumes (up to 15 volumes) at 0.3 m/s only yielded an additional 0.2–1.0 log (in the presence of a cultured biofilm) or 0.5–0.9 log (natural biofilm) removal from water (Figure 5C,D). The results show that flushing a pipe-loop system with a cultured or natural biofilm and with a low or high-speed yield comparable results. Flushing removes a large part of the microorganisms from the water, but removal is incomplete. Microorganisms bound to the biofilm are not removed or only to a limited extent. Although a natural and cultured biofilm can have different characteristics, their behavior in flushing experiments is comparable. Results of flushing experiments in a pipe-loop system with a cultured biofilm are thus predictive for results obtained with a natural biofilm.

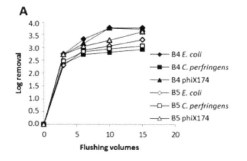
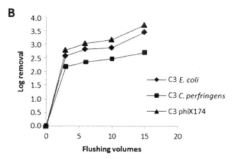
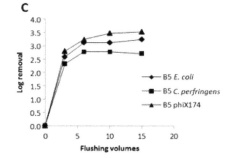

**Figure 6.** Flushing efficacy in log removal of microorganisms from the water and biofilm of a pipe with a cultured ((**A**,**C**); experiments B4 and B5) and natural biofilm ((**B**); experiment C3). In (**A**) (experiments B4–B5) and (**B**) (experiment C3) the pipe-loop system was flushed with 3 volumes (water) or 15 volumes (biofilm) at 0.3 m/s. In (**C**) (experiment B5) the pipe loop system was flushed with 15 volumes at 1.5 m/s. Samples were taken from the water after 3, 6, 10 and 15 volumes without stopping the flushing procedure.

### 3.8. Chlorine Disinfection

Flushing is effective in removing microorganisms from the water, but removal from the biofilm was shown to be very limited. Upon intrusion of (pathogenic) microorganisms in the drinking water distribution system during repairs or low-pressure transients or sites (reservoirs), binding of the microorganisms to the biofilm is likely to occur and, depending on the severity of the contamination, this may involve large numbers. To ensure clean and safe drinking water, intruded microorganisms that have attached to the biofilm have to be removed. Since this study shows that flushing has only very limited effect on the attached microorganisms, one of the possibilities is to combine flushing with shock chlorination, adding a high chlorine concentration (for example 10 mg free chlorine/L) to the drinking water and ensure that the network segment is treated with chlorination. The pipe segment is closed, without consumption and the chlorine is incubated for 24–48 h to kill off any microorganisms present in water and biofilm. The efficacy of chlorine on microorganisms in water has been studied before in laboratory situations but, to our knowledge, only very limited in a pilot distribution system [23]. The effect of chlorine on microorganisms that are present in or on the biofilm of the pipe wall is assumed to be less effective as the biofilm may protect the microorganisms. To determine the efficacy of chlorination, the flushing procedure was completed and subsequently chlorine was added to the pipe-loop system to a concentration of 10 mg/L and the chlorinated water was circulated for 24 h. The free chlorine concentration in the water was measured at regular intervals. After 24 h the chlorine concentration was 6.3–8.6 mg/L.

*E. coli* is very susceptible to chlorine and the log removal of culturable *E. coli* in water increases quickly to about 7 log removal (Figure 7). In experiments with a cultured biofilm, the number of *E. coli* in water is below the detection limit after 1 h of chlorine exposure (CT: 600 mg min/L), experiment B4). In the other two experiments the *E. coli* level in water is below the detection limit after 3.5 (experiment B1) or 6 (experiment B2) hours (CT: 2100–3600 mg min/L; white markings in the graphs of Figure 7). Similar die-off patterns can be seen when real-world pipes with a natural biofilm were incorporated. *E. coli* was not detectable in the water 1 (experiment C2) to 6 h (experiment C1 and C3) after dosage of chlorine, corresponding to 7–8 log removal. PhiX174 numbers in water, in the presence of a natural or cultured biofilm, were below the detection limit 1–6 h after addition of chlorine, corresponding to 7.5 log removal. *C. perfringens* spores were more resistant to chlorine and, in water, were still detected in three out of six experiments after 24 h of chlorination (CT: 14,400 mg min/L). In the biofilm die-off of all microorganisms was slower than in water. The initial decay upon chlorine dosage was quick, but levels off during the following 24 h. In the cultured biofilm *E. coli* was most sensitive to chlorine, followed by *C. perfringens* spores and phiX174. Only at 24 h the number of microorganisms sometimes dropped below the detection limit, but in most cases the microorganisms were still present in low numbers. In the natural biofilm die-off of microorganisms was slightly faster for *E. coli* and *C. perfringens*, but much faster for phiX174. Often the microorganisms cannot be detected after 1 h (*E. coli*) or 1–6 h (phiX174) of chlorination. *C. perfringens* remained present in two out of three experiments during the 24 h of chlorine exposure and were most resistant to chlorine. The initial decay upon chlorine dosage was quick, but levels off after 1 h and hardly any extra *C. perfringens* spores were inactivated. In most experiments the spores were still culturable after 24 h chlorine exposure. This suggests that a part of the *C. perfringens* spore population was quite sensitive to chlorine and were inactivated within one hour. The remaining population was more resistant and only a very limited effect of chlorine was visible.

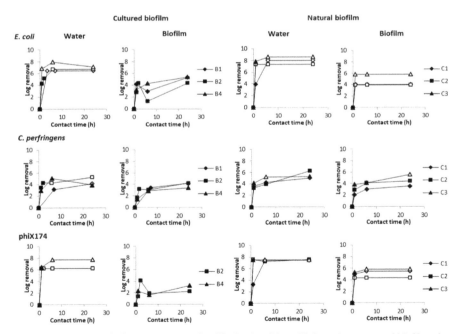

**Figure 7.** Log removal of microorganisms by chlorination (10 mg/L) from the water and biofilm of a cultured and natural biofilm in log removal. Markings in white indicate a result below the detection limit and is thus a '>value' in log removal.

It has long been known that chlorine disinfection is highly active against most bacteria [31,32] and viruses [33,34]. A CT-value around 1 mg min/L is often sufficient for 3–4 log reduction of these microorganisms. However, most protozoa and spores are highly resistant against chlorine and inactivation is limited [35,36]. In our experiments *C. perfringens* spores were most resistant to chlorine, but still 3–5 log was inactivated after 24 h incubation (CT: 14,400 mg min/L). In another study inactivation was lower: 1.7 log *C. perfringens* spores were inactivated upon incubation in a 5 mg/L chlorine solution for 24 h (CT: 7200 mg min/L) [37]. In the same study, *Cryptosporidium parvum* spores were completely resistant to the chlorine treatment. *Bacillus subtilis* spores were not inactivated during 180-min incubation in a 1–10 mg/L chlorine solution (CT: 180–1800 mg min/L) [22]. One of the reasons that cause the differences in inactivation kinetics may be the age of the spores. From UV-disinfection studies it is known that young spores are more vulnerable to UV disinfection then older spores [38]. It is possible that the same principle holds for chlorine disinfection, in that older spores are more resistant to chlorine than young spores. Disinfection studies in which the indicator viruses MS2 and phiX174 were incubated with 0.5 mg/L chlorine a >3.5 log reduction within 30 min for MS2 and 1 min for phiX174 was observed (CT: 15 and 0.5 mg min/L) [39]. In another study, applying 1–10 mg/L chlorine, a >5 log inactivation of free-suspended MS2 bacteriophage was achieved (CT: 15–20 mg min/L) [22]. PhiX174 was also quickly inactivated in our study (6 log inactivation, CT: 600 mg min/L). However, as inactivation was not measured in the first hour the results cannot be directly compared to literature.

For the same reason the inactivation kinetics of *E. coli* in our study (4–7 log, CT: 600 mg min/L) cannot be compared directly to literature in which *E. coli* seems apparently more vulnerable to chlorine (5–6 log inactivation within 5 min at 1 mg/L [22] or 4 log at 0.5 mg/L within 30 s) [32].

*3.9. Application to Real-Life Situation*

The flushing and disinfection experiments show that microorganisms in the water phase of a distribution system are readily removed by flushing. Although removal by flushing is not complete, in a real-life situation flushing is often probably sufficient as a contamination is likely to harbor less microorganisms than that were dosed in these experiments. A 2–3 log removal efficacy is sufficient in these cases. The tested flushing velocities (0.3 m/s or 1.5 m/s) yielded similar removal efficiencies. However, the microbial contamination was suspended in water and not bound to sediment particles which may make removal more difficult as a sufficient flow velocity is required to resuspend the particles in to the water phase. A previous study showed that lower flushing velocities (0.25–0.4 m/s) are sufficient for removal of sediment particles from a 100 mm pipe, however resuspension is likely to be more effective at 1.5 m/s compared to 0.3 m/s. Therefore, it remains recommendable to aim for a flushing velocity of 1.5 m/s. In situations in which 1.5 m/s cannot be reached due to practical reasons, flushing with 0.3 m/s can be used to remove the contamination from the water phase, but a less efficient removal of particles with attached microorganisms should be taken into account.

If a contamination enters the drinking water pipe and is not removed immediately, the microorganisms will have bound to the biofilm, and chlorine disinfection is required to inactivate these microorganisms. Chlorine disinfection should also be considered when the repair works have a high risk, for example when leaking sewage pipes are located close to or above the drinking water pipe line. The surrounding ground water or soil is likely to contain high numbers of microorganisms and even a contamination with a small amount of water or soil may introduce a large number of microorganisms. In these cases disinfection with a high chlorine concentration is required to inactivate the bound microorganisms and to ensure the microbiological safety of drinking water. Other options to remove, and thus clean the biofilm are available, for example, (ice) pigging during which an ice slurry is pushed through the pipe, removes the pipe wall biofilm and (nearly) all loose sediment particles. However, chlorination can be performed on a short notice, which is not possible for (ice) pigging.

Prevention of ingress of fecal contamination into the network remains pivotal in protecting the public health and requires strict adherence to hygienic working procedures. If the network becomes contaminated, protecting the public health remains the first priority. Therefore, not only flushing, and possibly chlorination, should be applied, but also a (preventive) boil water advice to the consumers.

*Water* **2019**, *11*, 903

Once the distribution system is contaminated it remains important to first localize and isolate the contamination source before performing flushing and chlorination.

## 4. Conclusions

Flushing and chlorine disinfection experiments in a pipe-loop system show that (1) sampling of water from the distribution system for microbiological monitoring can be performed 1 h after flushing, (2) small microbial contaminations in the water phase can be effectively flushed out of the system, (3) a contamination with sand was not completely removed from the system and present in the first few meters, (4) flushing with more than the prescribed three volume replacements removes only a limited number of extra microorganisms, (5) after high risk repair works or incidents shock chlorination is advised to ensure the microbiological safety of the drinking water as flushing is not efficient enough, (6) experiments with pipes with a cultured biofilm yield comparable results to pipes with a natural biofilm. For future experiments pipes with a cultured biofilm can be used, instead of pipes with a natural biofilm which have to be extracted from a distribution system by a drinking water company.

**Author Contributions:** Conceptualization, N.v.B., L.M.H. and G.M.; Investigation, N.v.B. and A.v.d.V.; Methodology, N.v.B., L.M.H. and A.v.d.V.; Visualization, N.v.B.; Writing—original draft, N.v.B.; Writing—review & editing, N.v.B. and G.M.

**Funding:** The research was funded by the Dutch drinking water companies through the Joint Research Program.

**Acknowledgments:** We would like to thank the Dutch drinking water companies, especially Jamal El Majjaoui (Dunea), Falco van Driel (Water Maatschappij Limburg) and Agata Donocik (Brabant Water), for arranging the excavation of the pipe segments from the real distribution network. The experiments would not have been possible without the technical assistance of the Laboratory of Microbiology of KWR.

## References

1. Karim, M.R.; Abbaszadegan, M.; LeChevallier, M. Potential for pathogen intrusion during pressure transients. *J. Am. Water Works Assoc.* **2003**, *95*, 134–146. [CrossRef]
2. Besner, M.C.; Broseus, R.; Lavoie, J.; Giovanni, G.D.; Payment, P.; Prevost, M. Pressure monitoring and characterization of external sources of contamination at the site of the payment drinking water epidemiological studies. *Environ. Sci. Technol.* **2010**, *44*, 269–277. [CrossRef] [PubMed]
3. Teunis, P.F.; Moe, C.L.; Liu, P.; Miller, S.E.; Lindesmith, L.; Baric, R.S.; Le Pendu, J.; Calderon, R.L. Norwalk virus: How infectious is it? *J. Med. Virol.* **2008**, *80*, 1468–1476. [CrossRef] [PubMed]
4. Messner, M.J.; Berger, P.; Nappier, S.P. Fractional poisson—A simple dose-response model for human norovirus. *Risk Anal.* **2014**, *34*, 1820–1829. [CrossRef] [PubMed]
5. Nygard, K.; Wahl, E.; Krogh, T.; Tveit, O.A.; Bohleng, E.; Tverdal, A.; Aavitsland, P. Breaks and maintenance work in the water distribution systems and gastrointestinal illness: A cohort study. *Int. J. Epidemiol.* **2007**, *36*, 873–880. [CrossRef] [PubMed]
6. Save-Soderbergh, M.; Bylund, J.; Malm, A.; Simonsson, M.; Toljander, J. Gastrointestinal illness linked to incidents in drinking water distribution networks in Sweden. *Water Res.* **2017**, *122*, 503–511. [CrossRef] [PubMed]
7. Payment, P.; Siemiatycki, J.; Richardson, L.; Renaud, G.; Franco, E.; Prevost, M. A prospective epidemiological study of gastrointestinal health effects due to the consumption of drinking water. *Int. J. Environ. Health Res.* **1997**, *7*, 5–31. [CrossRef]
8. Payment, P.; Richardson, L.; Siemiatycki, J.; Dewar, R.; Edwardes, M.; Franco, E. A randomized trial to evaluate the risk of gastrointestinal disease due to consumption of drinking water meeting current microbiological standards. *Am. J. Public Health* **1991**, *81*, 703–708. [CrossRef]
9. Nygard, K.; Andersson, Y.; Rottingen, J.A.; Svensson, A.; Lindback, J.; Kistemann, T.; Giesecke, J. Association between environmental risk factors and campylobacter infections in Sweden. *Epidemiol. Infect.* **2004**, *132*, 317–325. [CrossRef]

10. Hellard, M.E.; Sinclair, M.I.; Forbes, A.B.; Fairley, C.K. A randomized, blinded, controlled trial investigating the gastrointestinal health effects of drinking water quality. *Environ. Health Perspect.* **2001**, *109*, 773–778. [CrossRef]

11. Colford, J.M., Jr.; Wade, T.J.; Sandhu, S.K.; Wright, C.C.; Lee, S.; Shaw, S.; Fox, K.; Burns, S.; Benker, A.; Brookhart, M.A.; et al. A randomized, controlled trial of in-home drinking water intervention to reduce gastrointestinal illness. *Am. J. Epidemiol.* **2005**, *161*, 472–482. [CrossRef]

12. Ercumen, A.; Gruber, J.S.; Colford, J.M., Jr. Water distribution system deficiencies and gastrointestinal illness: A systematic review and meta-analysis. *Environ. Health Perspect.* **2014**, *122*, 651–660. [CrossRef] [PubMed]

13. Besner, M.C.; Prevost, M.; Regli, S. Assessing the public health risk of microbial intrusion events in distribution systems: Conceptual model, available data, and challenges. *Water Res.* **2011**, *45*, 961–979. [CrossRef] [PubMed]

14. Teunis, P.F.; Xu, M.; Fleming, K.K.; Yang, J.; Moe, C.L.; Lechevallier, M.W. Enteric virus infection risk from intrusion of sewage into a drinking water distribution network. *Environ. Sci. Technol.* **2010**, *44*, 8561–8566. [CrossRef]

15. Yang, J.; LeChevallier, M.W.; Teunis, P.F.; Xu, M. Managing risks from virus intrusion into water distribution systems due to pressure transients. *J. Water Health* **2011**, *9*, 291–305. [CrossRef]

16. Davis, M.J.; Janke, R. Development of a Probabilistic Timing Model for the Ingestion of Tap Water. *J. Water Resour. Plan. Manag.* **2009**, *135*, 397–405. [CrossRef]

17. Blokker, E.; Smeets, P.; Medema, G. QMRA in the Drinking Water Distribution System. *Procedia Eng.* **2014**, *89*, 151–159. [CrossRef]

18. Blokker, M.; Smeets, P.; Medema, G. Quantitative microbial risk assessment of repairs of the drinking water distribution system. *Microb. Risk Anal.* **2017**. [CrossRef]

19. Teunis, P.F.M.; Chappell, C.L.; Okhuysen, P.C. Cryptosporidium Dose-Response Studies: Variation Between Hosts. *Risk Anal.* **2002**, *22*, 475–485. [CrossRef]

20. Regli, S.; Rose, J.B.; Haas, C.N.; Gerba, C.P. Modeling the Risk From Giardia and Viruses in Drinking Water. *J. Am. Water Works Assoc.* **1991**, *83*, 76–84. [CrossRef]

21. Teunis, P.; van der Heijden, O.; van der Giessen, J.; Havelaar, A.H. *THE DOSE-RESPONSE RELATION in Human Volunteers for Gastro-Intestinal Pathogens*; Rijksinstituut Voor Volksgezondheid en Milieu RIVM: Bilthoven, The Netherlands, 1996.

22. Yang, J.; Schneider, O.D.; Jjemba, P.K.; Lechevallier, M.W. Microbial Risk Modeling for Main Breaks. *J. Am. Water Works Assoc.* **2015**, *107*, E97–E108. [CrossRef]

23. Krishnan, R.P.; Randall, P. *Pilot-Scale Tests and Systems Evaluation for the Containment, Treatment, and Decontamination of Selected Materials from T&E Building Pipe Loop Equipment*; U.S. Environmental Protection Agency: Washington, DC, USA, 2008.

24. McMinn, B.R.; Ashbolt, N.J.; Korajkic, A. Bacteriophages as indicators of faecal pollution and enteric virus removal. *Lett. Appl. Microbiol.* **2017**, *65*, 11–26. [CrossRef]

25. Wingender, J.; Flemming, H.C. Contamination potential of drinking water distribution network biofilms. *Water Sci. Technol.* **2004**, *49*, 277–286. [CrossRef]

26. Langmark, J.; Storey, M.V.; Ashbolt, N.J.; Stenstrom, T.A. Biofilms in an urban water distribution system: Measurement of biofilm biomass, pathogens and pathogen persistence within the Greater Stockholm Area, Sweden. *Water Sci. Technol.* **2005**, *52*, 181–189. [CrossRef]

27. Bauman, W.J.; Nocker, A.; Jones, W.L.; Camper, A.K. Retention of a model pathogen in a porous media biofilm. *Biofouling* **2009**, *25*, 229–240. [CrossRef]

28. Paris, T.; Skali-Lami, S.; Block, J.C. Probing young drinking water biofilms with hard and soft particles. *Water Res.* **2009**, *43*, 117–126. [CrossRef]

29. Helmi, K.; Skraber, S.; Gantzer, C.; Willame, R.; Hoffmann, L.; Cauchie, H.M. Interactions of Cryptosporidium parvum, Giardia lamblia, vaccinal poliovirus type 1, and bacteriophages phiX174 and MS2 with a drinking water biofilm and a wastewater biofilm. *Appl. Environ. Microbiol.* **2008**, *74*, 2079–2088. [CrossRef]

30. Mains, D.W. American Water Works Association. In *ANSI/AWWA C651-14*; American Water Works Association: New York, NY, USA, 2015. [CrossRef]

31. Blaser, M.J.; Smith, P.F.; Wang, W.L.; Hoff, J.C. Inactivation of Campylobacter jejuni by chlorine and monochloramine. *Appl. Environ. Microbiol.* **1986**, *51*, 307–311.

32. Zhao, T.; Doyle, M.P.; Zhao, P.; Blake, P.; Wu, F.M. Chlorine inactivation of *Escherichia coli* O157:H7 in water. *J. Food Prot.* **2001**, *64*, 1607–1609. [CrossRef]
33. Engelbrecht, R.S.; Weber, M.J.; Salter, B.L.; Schmidt, C.A. Comparative inactivation of viruses by chlorine. *Appl. Environ. Microbiol.* **1980**, *40*, 249–256.
34. Thurston-Enriquez, J.A.; Haas, C.N.; Jacangelo, J.; Gerba, C.P. Chlorine inactivation of adenovirus type 40 and feline calicivirus. *Appl. Environ. Microbiol.* **2003**, *69*, 3979–3985. [CrossRef] [PubMed]
35. Jarroll, E.L.; Bingham, A.K.; Meyer, E.A. Effect of chlorine on Giardia lamblia cyst viability. *Appl. Environ. Microbiol.* **1981**, *41*, 483–487.
36. Shields, J.M.; Hill, V.R.; Arrowood, M.J.; Beach, M.J. Inactivation of Cryptosporidium parvum under chlorinated recreational water conditions. *J. Water Health* **2008**, *6*, 513–520. [CrossRef] [PubMed]
37. Venczel, L.V.; Arrowood, M.; Hurd, M.; Sobsey, M.D. Inactivation of Cryptosporidium parvum Oocysts and Clostridium perfringens Spores by a Mixed-Oxidant Disinfectant and by Free Chlorine. *Appl. Environ. Microbiol.* **1997**, *63*, 4625. [PubMed]
38. Hijnen, W.A.; Beerendonk, E.F.; Medema, G.J. Inactivation credit of UV radiation for viruses, bacteria and protozoan (oo)cysts in water: A review. *Water Res.* **2006**, *40*, 3–22. [CrossRef]
39. Sobsey, M.D.; Fuji, T.; Shields, P.A. Inactivation of Hepatitis a Virus and Model Viruses in Water by Free Chlorine and Monochloramine. *Water Sci. Technol.* **1988**, *20*, 385–391. [CrossRef]

*Article*

# Managing Water Quality in Intermittent Supply Systems: The Case of Mukono Town, Uganda

**Takuya Sakomoto** [1] **and Edo Abraham** [3,*]

1   Department of Engineering, Course of Social Systems and Civil Engineering, Tottori University, Tottori 680-8552, Japan; tottori.b14t7024z@gmail.com
2   National Water and Sewerage Corporation, P.O. Box 7053, Kampala, Uganda; mlutaaya.ml@gmail.com
3   Department of Water Management, Delft University of Technology, 2628 CN Delft, The Netherlands
*   Correspondence: E.Abraham@tudelft.nl

Received: 31 January 2020; Accepted: 11 March 2020; Published: 13 March 2020

**Abstract:** Intermittent water supply networks risk microbial and chemical contamination through multiple mechanisms. In particular, in the cities of developing countries, where intrusion through leaky pipes are more prevalent and the sanitation systems coverage is low, contaminated water can be a public health hazard. Although countries using intermittent water supply systems aim to change to continuous water supply systems—for example, Kampala city is targeting to change to continuous water supply by 2025 through an expansion and rehabilitation of the pipe infrastructure—it is unlikely that this transition will happen soon because of rapid urbanisation and economic feasibility challenges. Therefore, water utilities need to find ways to supply safe drinking water using existing systems until gradually changing to a continuous supply system. This study describes solutions for improving water quality in Mukono town in Uganda through a combination of water quality monitoring (e.g., identifying potential intrusion hotspots into the pipeline using field measurements) and interventions (e.g., booster chlorination). In addition to measuring and analyses of multiple chemical and microbial water quality parameters, we used EPANET 2.0 to simulate the water quality dynamics in the transport pipeline to assess the impact of interventions.

**Keywords:** intermittent water supply; microbial contamination; drinking water quality modelling; sustainable development goals (SDG6)

## 1. Introduction

Intermittent water supply systems represent a range of water supply services that supply water to consumers for less than 24 h per day or not at sufficiently high pressures [1]. Such systems are typical in developing countries, with more than 1.3 billion people in at least 45 low- and middle-income countries reportedly receiving water through intermittent systems [2–4]. It is generally considered that intermittent water supply systems are not an ideal method of supply and do not constitute the best solution [5]. Where such systems are poorly maintained and have leaky pipe infrastructure, contamination can intrude into the water distribution system when pipes are at low or zero pressure. This could be through infrastructure deficiencies (e.g., holes and cracks in some pipes due to aging and deterioration) or break flows through cross connections (a plumbed connection between a potable water supply and a non-potable water source) and this could happen through events that are persistent or temporary [6–9]. In particular, contamination of pathogenic microorganisms such as *E. coli* in water cause various diseases [10]. However, intermittent water supply is being used by many water utilities to address water shortages (e.g., in drought conditions) or increasing demand, without considering long-term alternative solutions [1]. Although essential, transitioning from intermittent water supply systems to continuous water supply systems is often difficult for utilities [11,12] and needs to be done in a cost-effective way, combining improved operations with targeted capital works [1].

In Uganda Vision 2040 [13], the government aims "to improve the health, sanitation, hygiene, promote commercial and low consumption industrial setups, Government (sic) will construct and extend piped water supply and sanitation systems to all parts of the country". To accomplish this, the Ministry of Water outlines "improved water quality" and "reliable water quantity" as actionable targets for 2040 [14]. The planned transitions are even more ambitious for Kampala and target 2025 with the Kampala Water Project to rehabilitate the distribution network and extend water treatment capacity to meet all needs [14], in a bid to attain Sustainable Development Goal (SDG) 6, which focuses on sustainable access to clean water and sanitation. This means it will take up to 20 years to change all water supply systems in Uganda to new continuous systems. However, safe water quality is required under intermittent water supply systems until continuous supply goals are achieved.

In previous research, residual chlorine values in the Kampala water distribution system were investigated to assess the potential of recontamination [15]. Although Ecuru et al. [15] were only concerned with physicochemical water quality parameters (temperature, pH, turbidity, colour, ammonia, $Fe^{2+}$, and free chlorine), they were able to show low levels of residual chlorine compared to the minimum standard requirements. However, microbial water quality parameters were missing. In addition, interventions on how to maintain the level of residual chlorine in the water distribution systems, as well as parameters for other water pollutants were also not investigated.

In this manuscript, we assess multiple chemical and microbial water quality parameters in the Kampala water distribution system and examine the impact on water quality in Mukono town because of low pressure and leaky pipe infrastructure under intermittent water supply systems. The treatment works in Kampala use conventional chlorination as a disinfectant with the aim of maintaining disinfectant residuals throughout the supply network. We, therefore, measure residual chlorine levels at different parts of the water distribution system as an indicator for remaining disinfection potential and assess multiple water quality parameters, including microbial water quality (e.g., *E. coli*), to examine the water quality from the treatment plant to the very end of the network. The study is based on sample data collected at multiple sampling points between Ggaba II (water treatment plant) and Mukono town, via the centre of Kampala. This study considers pH, colour, turbidity, nitrate, nitrite, ammonia, sulphate, *E. coli*, total coliform, COD (Chemical Oxygen Demand) and chlorine because these parameters are related to water contamination by chemical and microbial pollutants that can cause health hazards [15]. Based on this assessment, we also propose booster chlorination in the supply reservoirs to reduce the microbial risk within the existing intermittent water supply system. By coupling the modelling of the distribution network hydraulics and chlorine decay processes in EPANET 2.0 [16], with residual chlorine levels from measurements, we propose feasible booster chlorine levels to achieve minimum standards for residual chlorine at all consumption nodes.

## 2. Materials and Methods

### 2.1. Current State of Site and Its Water Distribution System

Our study focuses on Mukono, a town in the north east of Kampala with a population of 664,300 (in 2018) [17]. Water is supplied to Mukono for an average of 20 to 24 h per day to all customers and sourced from Lake Victoria, as with most towns in Uganda [4]. Water pumped up from Lake Victoria is supplied to customers after a treatment process that includes coagulation, flocculation, sedimentation, gravel and sand filtration, disinfection with chlorine gas and pH adjustment by soda [4,18]. The design capacity of Ggaba II is 80,000 $m^3$/day with an average production of about 70,252 $m^3$/day [19]. Water is transmitted to the primary reservoirs of Muyenga, Seeta and Mukono (see Figure 1 and Table 1). The distribution and transmission pipes range in diameter from 50 to 900 millimetres.

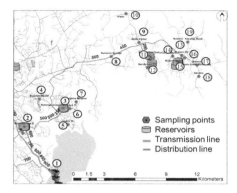

**Figure 1.** Kampala water distribution systems and pipelines from Ggaba II to Mukono town. The names of the numbered sites are shown in Table 1.

**Table 1.** The number and name of each sampling location as well as their respective elevation in the Kampala water distribution system (elevation data was acquired from the National Water and Sewerage Corporation (NWSC)'s GIS (Geographic Information System) data set).

| Number | Name of Place | Elevation (m) | Number | Name of Place | Elevation (m) |
|--------|---------------|---------------|--------|---------------|---------------|
| 1 | Ggaba II | 1126 | 11 | Seeta tank lower | 1170 |
| 2 | Muyunga Tank E | 1232 | 12 | Namilyango Trading center | 1196 |
| 3 | Mutungo reserver | 1228 | 13 | Mukono Health Center | 1168 |
| 4 | Bugorobi Market | 1172 | 14 | Mukono UCU | 1256 |
| 5 | Luzira Trading Center | 1146 | 15 | Mukono tank A+B | 1164 |
| 6 | Mutungo Trading center | 1168 | 16 | Mukono Market | 1240 |
| 7 | Butabika Hospital | 1168 | 17 | Mukono Wantoni | 1176 |
| 8 | Namanve Booster | 1134 | 18 | Mukono Dandira | 1160 |
| 9 | Seeta market | 1182 | 19 | Mukono Kayunga Road | 1172 |
| 10 | Bukerere Joggo | 1202 | | | |

## 2.2. Sampling Locations

This study was conducted on the Kampala water supply line from Ggaba II to Mukono town, via Kampala, in Uganda during January 2019. Kampala is the capital city of Uganda. This study focused on main points (19 sites) of the water supply system, such as the water treatment plant (Ggaba II), water storage tanks (supply reservoirs) and end user taps at various locations (a hospital, a market, a trading centre, a pump station and a university). The name of places, locations and elevations are shown in Figure 1 and listed in Table 1.

## 2.3. Choice of Water Quality Parameters

To assess water quality, twelve water parameters were selected for the analysis and simulation of the supply system (Table 2): free chlorine, *E. coli*, total coliform, COD, ammonia, nitrate, nitrite, sulphate, turbidity, pH, water age and water pressure. All samples were collected and tested within the standard 30-h holding time [20].

**Table 2.** Measurement kit and device to measure water quality of tap water.

| Measurement Items | Measurement Devices |
|-------------------|---------------------|
| *E. coli*, Total coliform | Compact Dry EC Nissui for Coliform and *E. coli* |
| Free chlorine, COD, Ammonia, Nitrate, Nitrite, Sulphate, Turbidity | Photometer PF-12$^{Plus}$ and NANOCOLOR tube test |

These parameters were chosen for multiple reasons. For example, changes in residual chlorine are linked with pollution by chemicals like ammonia and biofilm growth; chlorine reactions with corrosion products and COD could most likely explain low residual chlorine levels at the end of the network, where water age can be high [21]. The low residual chlorine levels in the Mukono reservoir and areas that follow are more likely due to this reason. Higher turbidity levels are often associated with higher levels of viruses, parasites and some bacteria [22]. There can be correlations between water age and chlorine levels [21]. Exposure to extreme pH values results in irritation to the eyes, skin, and mucous membranes [23]. Careful attention to pH control is necessary at all stages of water treatment to ensure satisfactory water purification and disinfection [24]. For effective disinfection with chlorine, the pH should preferably be less than eight and pH levels that are too low can affect the degree of corrosion of metals [23]. Maintaining continuous optimal water pressure in drinking water distribution systems can protect water from contamination as it flows to consumer taps [25,26]. Sulphate is one of the major dissolved components of rain. High concentrations of sulphates in drinking water can have a laxative effect [27]. Ammonia, nitrite and nitrate can cause algae growth [28] and can be indicators of pollution from agriculture and anthropogenic sources. Ammonia is an important indicator of faecal pollution [29]. Nitrite and nitrate have toxic effects for infants under three months of age and so require monitoring [30].

*2.4. Sampling Methods, Measurement Items and Analysis*

We followed recommended methods, also used by the National Water and Sewerage Corporation (NWSC), to collect samples. Samples for chemical analysis were collected in clean 500 ml plastic bottles. After flushing the taps for more than 10 seconds in order to clean the faucet, the sampling bottle was rinsed three times with the same water source as the sample before sampling. Collected samples were brought back to a laboratory using a cold storage case to keep them at low temperatures (at under 10 °C) to prevent bacterial growth during transport [20].

The measurement items and devices are listed in Table 2. Free chlorine was tested twice on-site using Photometer PF-12$^{Plus}$ (MACHEREY-NAGEL, Düren, Germany) and the NANOCOLOR tube test (MACHEREY-NAGEL, Düren, Germany) by following the procedure of NANOCOLOR tube test. The samples for *E. coli* and total coliform were tested by incubating for 24 h at 35 °C $\pm$ 2 in an incubator by Compact Dry EC Nissui for coliform and *E. coli* detection (Nissui, Tokyo, Japan) by following the protocol for Compact Dry EC Nissui for coliform and *E. coli* detection. Nitrite, nitrate, sulphate, ammonia and turbidity were tested twice for each sample by Photometer PF-12$^{Plus}$ and the NANOCOLOR tube test in the laboratory by following the procedure of the NANOCOLOR tube test, and were validated with pH data from January 2019 using records from internal GIS and measurement databases maintained by the operations team at NWSC. Water age and water pressure were assessed using EPANET simulations. Measurement ranges for parameters were selected to include recommended standards by WHO [24], the Ministry of the Environment Government of Japan [31], or as in Wang et al. [32] and Kumpel et al. [8] (See Table 3).

**Table 3.** Measured parameters of water quality, the standards of water quality and range of photometer used.

| Measurement Items | Measurement Standard | Range of Photometer |
|---|---|---|
| Free chlorine | 0.5 mg/L (0.2–1.0) [24] | - |
| *E. coli* | 0 (CFU/100mL) [24] | - |
| Total coliform | 0 (CFU/100mL) [24] | - |
| COD | ≤5 mg/L [31] | 3 to 150 mg/L |
| Ammonia | ≤1.5 mg/L [24] | 0.04 to 2.30 mg/L |
| Nitrite | ≤3 mg/L [24] | 0.3 to 22.0 mg/L |
| Nitrate | ≤50 mg/L [24] | 0.1 to 4.0 mg/L |
| Sulphate | ≤250 mg/L [24] | 40 to 400 mg/L |
| Turbidity | ≤5 NTU [24] | 1 to 1000 NTU |
| pH | $6 \leq pH \leq 9$ [24] | - |
| Water age | ≤5.7 days (136.8 h) [32] | - |
| Water pressure | ≥17 psi [8] | - |

*2.5. Water Quality Modelling*

EPANET 2.0 is a computer program that can perform extended period simulation of hydraulic and water quality behaviour within pressurised pipe networks with pipes, nodes (pipe junctions), pumps, valves and storage tanks or reservoirs [16].

This paper used EPANET 2.0 to simulate chlorine decay, water age and pressure levels in the distribution network modelled. The pipe length and diameter data of the distribution network are shown in Figure 1 and presented in detail in Table 4. The considered part of the distribution network consists of five different pipe diameters and different lengths per diameter, which were all derived from the GIS dataset maintained by the National Water and Sewage Corporation (NWSC). Pipe roughness (Table 4) was derived using recommend values by Swierzawski (2000) for the material types and age [33], which are assumed to be the same for all pipe sections here. Future work can consider calibrating the roughness of the pipes, which may be different as thw pipes age. Three typical demand patterns in Kampala water distribution systems were generated to model the end-user demand for water in this study (Figure 2) [18]. Each pattern was applied based on the Republic of Uganda Ministry of water and environment's manual on demand patterns for Ugandan urban water use (2013), also shown in Table 5 [18]. EPANET was run to simulate water quality parameters for 168 h (i.e., 7 days).

**Table 4.** Pipe length, diameter and roughness of water distribution lines from Ggaba II to Mukono. The name of the numbered sites or nodes is as shown in Table 1.

| Pipe Name (from node No. x to y) | Pipe Length (m) | Pipe Diameter (mm) | Pipe Roughness |
|---|---|---|---|
| 1 to 2 | 5260 | 800 | 140 |
| 2 to 3 | 4210 | 500 | 140 |
| 3 to 4 | 2700 | 100 | 140 |
| 3 to 5 | 1740 | 100 | 140 |
| 3 to 6 | 660 | 100 | 140 |
| 6 to 7 | 1340 | 100 | 140 |
| 3 to 8 | 11700 | 500 | 140 |
| 8 to 11 | 4710 | 400 | 140 |
| 11 to 9 | 1680 | 200 | 140 |
| 11 to 12 | 1230 | 100 | 140 |
| 9 to 10 | 3260 | 100 | 140 |
| 9 to15 | 6920 | 400 | 140 |
| 15 to 13 | 1220 | 100 | 140 |
| 15 to 14 | 710 | 100 | 140 |
| 15 to 16 | 1530 | 100 | 140 |
| 16 to 17 | 1100 | 100 | 140 |
| 17 to 18 | 860 | 100 | 140 |
| 13 to 16 | 1240 | 100 | 140 |
| 13 to 19 | 1100 | 100 | 140 |

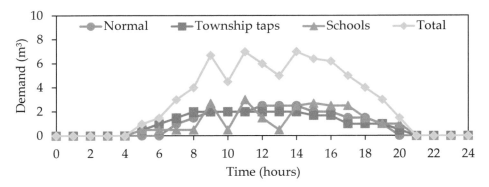

**Figure 2.** Typical weekday consumption patterns (normal households aggregated, township taps, schools and total of the three patterns) in Kampala water distribution systems [18].

**Table 5.** Demand pattern for each sampling points.

| No. | Pattern | No. | Pattern |
|-----|---------|-----|---------|
| 1 | - | 11 | - |
| 2 | - | 12 | Township taps |
| 3 | - | 13 | Normal |
| 4 | Township taps | 14 | School |
| 5 | Township taps | 15 | - |
| 6 | Township taps | 16 | Township taps |
| 7 | Township taps | 17 | Normal |
| 8 | - | 18 | Normal |
| 9 | Township taps | 19 | Normal |
| 10 | Normal | | |

2.5.1. Headloss in Pipes (Hazen–Williams Formula)

In this study, the Hazen–Williams formula was used to model frictional losses in the hydraulic model. The headloss across a pipe, per 100 feet of pipe, is given as:

$$Pd = \frac{4.52 \times Q^{1.85}}{C_{hw}^{1.85} \times d^{4.87}} \tag{1}$$

where $Q$ = flow rate (gpm or Lpm), $C_{hw}$. = roughness coefficient, dimensionless, $d$ = inside pipe diameter, in mm. The total head drop in the system (psi) as a function of the pipe length $L$ (ft) is given as [16]:

$$Td = 0,002082 \, L \times \frac{100^{1.85}}{C} \times \frac{Q^{1.85}}{d^{1.8655}} \tag{2}$$

with velocity ($V$), flow rate ($Q$) and friction headloss ($f$) are computed as follows:

$$V = 1.318 \times C_{hw} \times R^{0.63} \times S^{0.54} \tag{3}$$

$$Q = 0.849 \times C_{hw} \times A \times R^{0.63} \times S^{0.54} \tag{4}$$

$$f = \frac{6.05 \times Q^{1.85}}{C_{hw}^{1.85} \times d^{4.78}}$$

where $S$ = the slope of the energy line (head loss per length of pipe, unitless), $f$ = friction head loss in ft. hd./100 ft. of pipe (m per 100m), $Pd$ = pressure drop (psi/100 feet of pipe), $R$ = hydraulic radius, feet (m), $V$ = velocity (feet per second), and $A$ = cross section area, in (mm$^2$).

2.5.2. Chlorine Decay Kinetics

There are two main reasons why residual chlorine decays through reactive processes [34,35]. One reason can be external contamination, which happens often during pipeline breaks and maintenance operations, and the second is the decay with time through reaction with natural organic matter in the bulk water. The chlorine decay model [34–37] that is included in EPANET 2.0 [16] and used within our water quality analysis accounts for both bulk and wall reactions, with a limited growth on the ultimate concentration of the decaying substance, as follows:

$$r = k_b(C_L - C)C^{(n-1)}. \tag{5}$$

where $r$ is the rate of reaction (mass/volume/time), $kb$ is the reaction constant, $C$ is the reactant concentration (mass/volume), $n$ is the reaction order, and $C_L$ is the limiting concentration. Chlorine reactive decay in the bulk flow within the pipe is adequately modelled by a simple first-order reaction ($n = 1$, $k_b = 0.849$/s, $C_L = 0$ are used here [16]). The rate of the reactant reactions that happen at pipe walls are related to the reactant's concentration in the bulk, modelled as:

$$r_2 = (A/V)k_w C^n \tag{6}$$

where $r_2$ is the rate of reaction (mass/volume/time), $A/V$ is the surface area per unit volume within the pipe which equals (4/pipe diameter), $C$ is the chlorine concentration (mass/volume), $n$ is the kinetics order (1 here), and $k_w$ is the wall reaction rate coefficient (length/time for $n$=1 and mass/area/time for $n$=0). The EPANET simulation makes automatic adjustments to account for mass transfer between the bulk flow and the wall, based on the molecular diffusivity of the reactant under study and the Reynolds number of the flow. In the case of zero-order kinetics, which are recommended by the program manual [16], the wall reaction rate cannot be greater than the mass transfer rate, resulting in:

$$r_2 = -MIN(k_w, k_f C)(2/R), \tag{7}$$

where $k_f$ is the mass transfer coefficient (length/time), $C$ is the chlorine concentration (mass/volume), and $R$ is the pipe radius (length).

Water age was also calculated by EPANET 2.0 and was used to assess the age of each parcel of water in the system, where we have only one source at the treatment plant and the dynamics of service reservoirs (with a complete mixing model, i.e., the MIXED setting for tanks [16]) was employed. A "setpoint booster source" was used to fix the concentration of outflows leaving booster nodes, where the concentration of all inflow reaching the booster nodes was below the outflow chlorine concentration that was set. In this manuscript, Microsoft Office Excel (2016) was used to do the statistical analysis and visualize the results. Error bars were created using standard deviation P function in Microsoft Office Excel (2016).

## 3. Results and Discussion

Free chlorine levels measured for samples of the distribution system did not meet the WHO standard of ≥0.2 mg/L [24] at 15 sites. *E. coli* was detected at 11 sites (other bacteria were not detected). Although the average COD level did not meet the standard of ≤5 mg/L [31] at six locations, there were an additional four sites where one of the samples exceeded this threshold, i.e., locations three, four, nine and 15 in Figure 3c. Therefore, with more measurements, it is possible that more locations may have higher COD values.

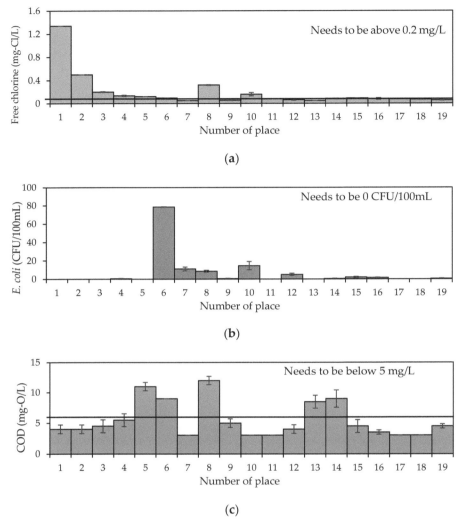

**Figure 3.** Free chlorine (**a**), *E. coli* (**b**) and COD (**c**) measurements from the Kampala water distribution system. The place numbers are as shown in Table 1 (n = 2). The error bars show the standard deviation of duplicate samples.

The average pH value in the water supply system was 7.1, the lowest was 6.8 (at Namanve booster) and the highest at 7.4 in Mukono market, which all meet the WHO standard set in [24]. Sulphate was detected at seven locations, but all below the maximum set by the WHO standard of ≤250 mg/L [24] (See also Table 6). Ammonia, nitrite, nitrate and turbidity were found to be below measurable limits by the photometer, which measured concentrations well below WHO [24] standard limits (ammonia: ≤1.5 mg/L, nitrite: ≤3 mg/L, nitrate: ≤50 mg/L, turbidity: ≤5 NTU) (see Table 5). Therefore, rainwater runoff and wastewater, such as industrial effluent and sewage, are not suspected of intruding in the drinking water distribution pipes based on these results [27–29].

**Table 6.** Ammonia, nitrate, nitrite, sulphate and turbidity from Kampala water distribution systems. Ammonia, nitrate and sulphate are shown in one item as they were not detected (or not found to be below measurable limits by the photometer) in two different measurements; these are indicated with a dash. The name of the numbered sites is as shown in Table 1.

| No. | Ammonia (mg N/L) | Nitrite (mg N/L) | Nitrate (mg N/L) | Sulphate (1st, 2nd) (mg S/L) | Turbidity (NTU) |
|---|---|---|---|---|---|
| 1 | - | - | - | - | - |
| 2 | - | - | - | - | - |
| 3 | - | - | - | - | - |
| 4 | - | - | - | 60, 40 | - |
| 5 | - | - | - | - | - |
| 6 | - | - | - | - | - |
| 7 | - | - | - | - | - |
| 8 | - | - | - | 54, 40 | - |
| 9 | - | - | - | - | - |
| 10 | - | - | - | 46, 47 | - |
| 11 | - | - | - | 48, 40 | - |
| 12 | - | - | - | - | - |
| 13 | - | - | - | - | - |
| 14 | - | - | - | - | - |
| 15 | - | - | - | - | - |
| 16 | - | - | - | 40, 48 | - |
| 17 | - | - | - | 73, 40 | - |
| 18 | - | - | - | - | - |
| 19 | - | - | - | 42, 46 | - |

The water age at most sample locations was estimated to be less than 20 h; the water age at Mukono Health Centre was 72.3 h (Figures 4 and 5), which is the highest value in this system, but is also well within the standard limit [32]. Seeta market, Namilyango TC (Trading Centre) and UCU (Uganda Christian University) Mukono (i.e., nodes numbered nine, 12 and 14 respectively) did not meet the minimum pressure head requirements of 17 psi at all times [8] (see Figure 6). At low pressures in the pipe, such as in UCU Mukono and Namilyango TC, contaminants could enter from the outside of the pipe.

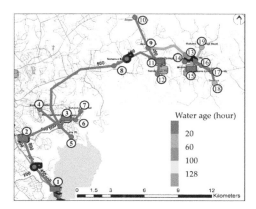

**Figure 4.** Water age in Kampala water distribution system. The overlaid colour of the node indicates water age after 130 h when the chlorine dynamics had stabilised with diurnal variations. The names of the numbered sites are shown in Table 1.

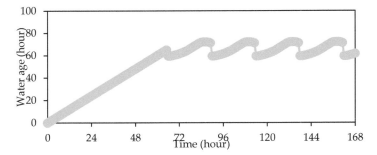

**Figure 5.** Water age at Mukono Health Centre (No. 13), the highest value in the system.

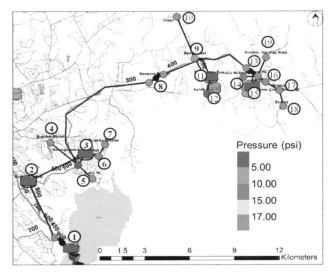

**Figure 6.** Kampala water distribution system and pipeline from Ggaba II to Mukono. The name of the numbered sites is shown in Table 1. The overlaid colour of nodes indicates water pressure.

Chlorine, *E. coli* and COD did not meet their respective standards at many locations. Using EPANET simulations, it was found that it is possible to maintain the free chlorine in the entire system above 0.7 mg/L at all times by adding chlorine, so that its concentration in the service tanks becomes sufficiently high (1.0 mg/L or more, see Figures 7 and 8). Since COD was below standard values at the exit of the water treatment plant, there is the possibility that organic matter entered from outside of the pipes due to low pressure, causing the increased COD at the sampling points, or the detachment of biofilm from pipe walls when hydraulic regimes/pressures change abruptly in the charging and discharging of pipes. If intrusion is a main culprit in COD increases, it is necessary to replace leaky pipes and joints that leak and improve operations that exasperate pipe damage and assess the impact using monitoring (e.g., COD and residual free chlorine). Free chlorine is sensitive to contaminants and its degradation can be used as an indicator of intrusion [38]. UCU Mukono and Namilyango TC, where pressure was low, did not meet free chlorine standards. Moreover, even at locations where minimum pressure standards were met, free chlorine levels did not meet the WHO standard. When comparing leaky sites and residual free chlorine along the network, we found that residual chlorine decreased after passing through Muyenga Tank E, where there are the most leaks reported (Figure 9). Similarly, cases of *E. coli* were mostly detected after Muyenga tank E. Therefore, intrusion from outside the pipes can be considered as probable cause. From these results, aging pipes are considered a main cause of water quality degradation in Mukono town. It has previously been reported that the large leakage levels, especially in the centre of Kampala as shown in Figure 9, are attributed to high rates of water theft, illegal connections, bursts and leakages [15,39]. Therefore, future work should explore the same parameters in the whole system, including a dense network of sampling in Kampala to get a better picture of risks associated to intrusion. Interventions in asset replacement can also be targeted to improve residual disinfection (e.g., plastic pipes, where free chlorine is more effective in preventing regrowth than in iron pipes [15]).

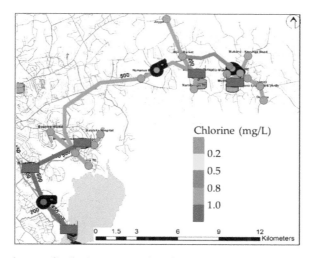

**Figure 7.** Kampala water distribution system and pipeline from Ggaba II to Mukono. The name of the numbered sites is shown on the Table 1. The overlaid colour of pipelines indicates free chlorine levels after 24 h adding chlorine at reservoirs.

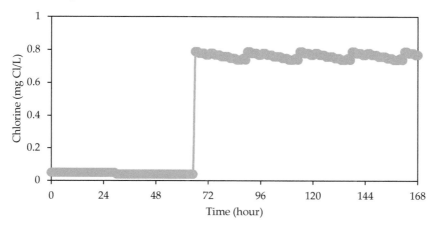

**Figure 8.** Free chlorine value at Mukono Health Centre (No. 13), lowest value for the system.

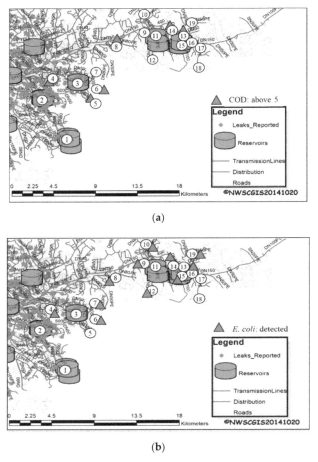

(a)

(b)

**Figure 9.** Kampala water distribution system and pipeline from Ggaba II to Mukono with leak occurrences (green circles) overlaid with *E. coli* measurements (**a**) and COD measurements (**b**). The name of the numbered sites is shown on the Table 1.

Although intrusion is our main suspect, it has also long been known that coliform regrowth (including faecal coliforms) is associated with biofilm regrowth in distribution systems [40,41]. High levels of coliform can be detected as a result of biofilm, even if very low levels are detected in effluent from the treatment plant. Since the residual chlorine levels of our case study were low anyway, it is possible that biofilms could play a part in bacterial regrowth. However, there were no known hydraulic disturbances and transients during the sampling campaign that imply the significant dislocation of biofilm and the very high levels of *E. coli* detected. Future work should consider the analysis of biofilm through grab sampling and fire hydrant experiments at a denser network of locations, taking into account the different pipe materials and ages. For example, heterotrophic plate counts (HPC) could be used as an indicator for the expanding of biofilms [21].

Although boosting the network with residual chlorine as a disinfectant is widely used, as proposed in our intervention, there are potential concerns that have to be considered in its use in addition to the capacity of residual chlorine to inactivate pathogens that gain entry to the network. Firstly, elevated levels of residual chlorine are associated with the unacceptable taste and odour of the drinking water for customers. However, our suggested intervention results in free chlorine levels were well below 1mg/L in the system, which is generally a range with no felt odour [42]. The case detection levels also vary widely between people, and can be based on the repeated experience of the taste [25]. Moreover, even if the dose needed to be high enough to result in a felt odour, it can be argued that safe water for the people (i.e., securing public and individual health) is more important than odour or taste. On the other hand, it is usually a low level of disinfectant that results in the development of discolouration, bad tastes and odours produced by biofilm. If such a case of felt odour is a concern in parts of the network, the utility should effectively communicate this to users for acceptance.

Another alternative is the use of chloramines instead of free chlorine as a residual disinfectant. This has been shown to perform better in the stability of residual concentrations throughout the distribution system, lower trihalomethane (THM) concentrations, better control of coliforms and heterotrophic plate count (HPC) bacteria [42]. It is also reported that chloramines result in less detectable taste and odour at much higher concentrations compared to free chlorine [25]. Although beyond the scope of this manuscript, the boost chlorination could also be optimised spatially and in time so that no part of the network has very high levels of residual chlorine. For example, common target concentrations for free chlorine in Europe are ~0.3 mg/L at the tap [25].

## 4. Conclusions

In this study, the water quality in the Kampala water distribution system was assessed from Ggaba II to Mukono town via Kampala and solutions were proposed to safeguard water quality. Using experimental data and EPANET modelling, boosting chlorine at tanks was proposed so that free chlorine concentration is at least 0.7 mg/L at all locations, to maintain residual disinfection and prevent contamination by *E. coli* and other biohazards. It is, therefore, possible to decrease intrusion in the network and increase the safety of water provided to consumers in Mukono town through a combination of water quality monitoring and booster disinfectant addition. Moreover, by showing a potential correlation between the water quality degradation and location of leaky pipes and low pressures, it was also recommended that preferentially upgrading leaky and damaged pipes should be done.

A further action for research would be to test the proposed intervention within the real distribution system in order to demonstrate its efficacy practically. Moreover, water quality data, which is collected continuously or through repetitive grab sampling, is needed to characterise disinfectant residual contaminations and indicator bacteria concentrations throughout the supply cycle, including during first flush [43]. As such, the approach can be applied in other similar systems using intermittent water supply to improve the water quality while the system transitions to continuous supply over many years. In future, inline disinfectant addition locations could also be optimised for a more uniform chlorine distribution [44,45].

**Author Contributions:** Conceptualization, T.S. and E.A.; investigation, T.S.; methodology, T.S., M.L. and E.A.; software, T.S. and E.A.; funding acquisition, T.S. and E.A.; writing—original draft preparation, T.S.; writing—review and editing, T.S., M.L. and E.A.; visualization, T.S.; supervision, E.A. All authors have read and agreed to the published version of the manuscript.

**Funding:** The visit of Takuya Sakomoto in Delft and transport to Uganda was sponsored by the Japan Public–Private Partnership Student Study Abroad TOBITATE! Young Ambassador Program (members from MEXT (Ministry of Education, Culture, Sports, Science and Technology), JASSO (Japan Student Services Organization), private sectors and universities) and the experimental analysis was partly financed by the TU Delft Global Initiative, a program of Delft University of Technology to boost science and technology for global development.

**Acknowledgments:** The authors are grateful to Christipher Kanyesigye and Enos Malambula of NWSC for providing laboratory space, support in the sampling process and the network data.

**Conflicts of Interest:** The authors declare no conflict of interest.

## References

1. Charalambous, B.; Liemberger, R. *Dealing with the Complex Interrelation of Intermittent Supply and Water Losses*; International Water Association: London, UK, 2017; Available online: https://www.iwapublishing.com/books/9781780407067/dealing-complex-interrelation-intermittent-supply-and-water-losses (accessed on 12 January 2019).

2. Van den Berg, C.; Danilenko, A. *The IBNET Water Supply and Sanitation Performance Blue Book*; The World Bank: Washington, DC, USA, 2011; Available online: http://documents.worldbank.org/curated/en/420251468325154730/The-IBNET-water-supply-and-sanitation-performance-blue-book (accessed on 22 February 2019).

3. World Health Organization; UNICEF. *Global Water Supply and Sanitation Assessment 2000 Report*; World Health Organization: Geneva, Switzerland; UNICEF: New York, NY, USA, 2000; Available online: https://www.who.int/water_sanitation_health/monitoring/jmp2000.pdf (accessed on 12 January 2019).

4. Government of Uganda Ministry of Water and Environment. *Water and Sanitation Sector Performance Report 2006*; Government of Uganda: Kampala, Uganda, 2006. Available online: https://docs.google.com/file/d/0BwhPKU71ZwQDSDc3eTdXMlpLV1U/edit (accessed on 22 June 2019).

5. Intermittent Water Supply—A Paradigm Shift is Imperative. Available online: https://iwa-network.org/intermittent-water-supply-a-paradigm-shift-is-imperative/ (accessed on 25 June 2019).

6. Vairavamoorthy, K.; Gorantiwar, S.D.; Mohan, S. Intermittent water supply under water scarcity situations. *Water Int.* **2007**, *32*, 121–132. [CrossRef]

7. Barwick, R.S.; Uzicanin, A.; Lareau, S.; Malakmadze, N.; Imnadze, P.; Iosava, M.; Ninashvili, N.; Wilson, M.; Hightower, A.W.; Johnston, S.; et al. Outbreak of Amebiasis in Tbilisi, republic of Georgia, 1998. *Am. Trop. Med. Hyg.* **2002**, *67*, 623–631. [CrossRef]

8. Kumpel, E.; Nelson, K.L. Mechanisms affecting water quality in an intermittent piped water supply. *Environ. Sci. Technol.* **2014**, *48*, 2766–2775. [CrossRef] [PubMed]

9. Mutikanga, H.E.; Sharma, S.K.; Vairavamoorthy, K. Assessment of apparent losses in urban water systems. *Water Environ. J.* **2011**, *25*, 327–335. [CrossRef]

10. Haward, G.; Pedley, S.; Tibatemwa, S. Quantitative microbial risk assessment to estimate health risks attributable to water supply: Can the technique be applied in developing countries with limited data? *J. Water Health* **2006**, *4*, 49–65. [CrossRef]

11. IWA Connect. Available online: https://iwa-connect.org/group/intermittent-water-supply-iws/about?view=public (accessed on 26 June 2019).

12. Klingel, P.; Nestmann, F. From intermittent to continuous water distribution: A proposed conceptual approach and a case study of Béni Abbès (Algeria). *Urban Water J.* **2014**, *11*, 240–251. [CrossRef]

13. National Planning Authority (NPA), Government of Uganda. *Uganda Vision 2040*; NPA: Kampala, Uganda, 2013. Available online: https://www.greengrowthknowledge.org/sites/default/files/downloads/policy-database/UGANDA%29%20Vision%202040.pdf (accessed on 8 January 2019).

14. The Republic of Uganda, Ministry of Water and Environment. *Framework and Guidelines for Water Source Protection Volume 2: Guidelines for Protecting Water Sources for Piped Water Supply Systems*; The Republic of Uganda Ministry of Water and Environment: Kampala, Uganda, 2013. Available online: https://www.mwe.go.ug/sites/default/files/library/Vol.%202%20-%20Guidelines%20for%20Protecting%20Piped%20Water%20Sources%20-%20FINAL.pdf (accessed on 10 January 2019).

15. Ecuru, J.; Okumu, O.J.; Okurut, O.T. Monitoring residual chlorine decay and coliform contamination in water distribution network of Kampala, Uganda. *Afr. J. Online* **2011**, *15*, 167–173. [CrossRef]

16. Rossman, L. *EPANET 2 User's Manual*; Risk Reduction Engineering Laboratory, U.S. Environmental Protection Agency: Cincinnati, OH, USA, 2000.

17. Uganda Bureau of Statistics Population and Censuses, Population Projections 2018. Available online: https://www.ubos.org/explore-statistics/20/ (accessed on 30 June 2019).

18. The Republic of Uganda Ministry of water and environment. *Water Supply Design Manual Second Edition*; The Republic of Uganda Ministry of Water and Environment: Kampala, Uganda, 2013. Available online: https://www.mwe.go.ug/sites/default/files/library/Water%20Supply%20Design%20Manual%20v.v1.1.pdf (accessed on 12 January 2019).

19. Kalibbala, H.M.; Nalubega, M.; Wahlberg, O.; Hultman, B. Performance Evaluation of Drinking Water Treatment Plants in Kampala—Case of Ggaba II. In Proceedings of the 32nd WEDC International Conference, Colombo, Sri Lanka, 13–17 November 2006; pp. 373–376.

20. Environmental Protection Agency. *Quick Guide to Drinking Water Sample Collection*; Environmental Protection Agency: Washington, DC, USA, 2016. Available online: https://www.epa.gov/sites/production/files/2015-11/documents/drinking_water_sample_collection.pdf (accessed on 22 February 2019).

21. Blokker, M.; Furnass, W.; Machell, J.; Mounce, S.; Schaap, P.G.; Boxall, J. Boxall. Relating Water Quality and age in drinking water distribution systems using self-organising maps. *Environments* **2016**, *3*, 10. [CrossRef]

22. Minnesota Pollution Control Agency. *Turbidity: Description, Impact on Water Quality, Sources, Measures—A General Overview*; Minnesota Pollution Control Agency: St Paul, MN, USA, 2008. Available online: https://www.pca.state.mn.us/sites/default/files/wq-iw3-21.pdf (accessed on 20 February 2019).

23. World Health Organization. *pH in Drinking-Water Background Document for Development of WHO Guidelines for Drinking-Water Quality*; World Health Organization: Geneva, Switzerland, 2014.

24. Cotruvo, J.A. 2017 WHO guidelines for drinking water quality: First addendum to the fourth edition. *J. Am. Water Works Assoc.* **2017**, *109*, 44–51. [CrossRef]

25. Ainsworth, R. *Safe Piped Water: Managing Microbial Water Quality in Piped Distribution Systems*; World Health Organization: Geneva, Switzerland, 2004; Available online: https://apps.who.int/iris/handle/10665/42785 (accessed on 12 January 2019).

26. Geldreich, E. *Microbial Quality of Water Supply in Distribution Systems*; CRC Lewis Publishers: Boca Raton, FL, USA, 1996.

27. William, D.H.; Robert, S.S.; Elston, S.; Sharon, C.M.; Marjorie, G.B.; Barbara, G.S.; Susan, N.P. Intestinal effects of sulphate in drinking water on normal human subjects. *Dig. Dis. Sci.* **1997**, *42*, 1055–1061. [CrossRef]

28. Prest, E.I.; Hammes, F.; van Loosdrecht, M.C.; Vrouwenvelder, J.S. Biological stability of drinking water: Controlling factors, methods, and challenges. *Front. Microbiol.* **2016**, *7*, 45. [CrossRef] [PubMed]

29. World Health Organization. *Ammonia in drinking-Water Background Document for Development of WHO Guidelines for Drinking-Water Quality*; World Health Organization: Geneva, Switzerland, 2001; Available online: https://www.who.int/water_sanitation_health/water-quality/guidelines/chemicals/ammonia.pdf?ua=1 (accessed on 12 January 2019).

30. World Health Organization. *Nitrate and Nitrite in Drinking-Water Background Document for Development of WHO Guidelines for Drinking-Water Quality*; World Health Organization: Geneva, Switzerland, 2016; Available online: https://apps.who.int/iris/handle/10665/75380 (accessed on 12 January 2019).

31. Ministry of the Environment Government of Japan. Available online: https://www.env.go.jp/en/index.html (accessed on 12 January 2019).

32. Wang, H.; Masters, S.; Hong, Y.; Stallings, J.; Falkinham, J.O., III; Edwards, M.A.; Pruden, A. Effect of disinfectant, water age, and pipe material on occurrence and persistence of Legionella, mycobacteria, Pseudomonas aeruginosa, and two amoebas. *Environ. Sci. Technol.* **2012**, *46*, 11566–11574. [CrossRef] [PubMed]

33. Tadeusz, J. Swierzawski. *Piping Handbook 7th Edition, Flow and fluids Chapter B8*; MCGRAW-HILL: New York, NY, USA, 2000; Available online: https://engineeringdocu.files.wordpress.com/2011/12/piping-handbook.pdf (accessed on 23 February 2019).

34. Ozdemir, O.N.; Ucak, A. Simulation of chlorine decay in drinking-water distribution systems. *J. Environ. Eng.* **2002**, *128*, 31–39. [CrossRef]

35. Vieira, P.; Coelho, S.T.; Loureiro, D. Accounting for the influence of initial chlorine concentration, TOC, iron and temperature when modeling chlorine decay in water supply. *J. Water Supply Res. Technol. AQUA* **2004**, *53*, 453–467. [CrossRef]

36. Al-Jasser, A.O. Chlorine Decay in Drinking-Water Transmission and distribution systems: Pipe service age effect. *Water Res.* **2007**, *41*, 387–396. [CrossRef]

37. Housing and Building National Research Centre. *Egyptian Code for Design and Implementation of Pipeline Networks for Drinking Water and Sanitation*, 10th ed.; Housing and Building National Research Centre: Cairo, Egypt, 2007.

38. Helbling, D.E.; Van Briesen, J.M. Modeling residual chlorine response to a microbial contamination event in drinking water distribution systems. *J. Environ. Eng.* **2009**, *135*, 918–927. [CrossRef]

39.  The Republic of Uganda, Ministry of Water and Environment. *Uganda Water and Environment Sector Performance Report 2017*; The Republic of Uganda Ministry of Water and Environment: Kampala, Uganda, 2017. Available online: https://www.mwe.go.ug/sites/default/files/library/SPR%202017%20Final.pdf (accessed on 20 February 2020).
40.  LeChevallier, M.W.; Babcock, T.M.; Lee, R.G. Examination and characterization of distribution system biofilms. *Appl. Environ. Microbiol.* **1987**, *53*, 2714–2724. [CrossRef]
41.  Flemming, H.-C.; Percival, S.; Walker, J.T. Contamination potential of biofilms in water distribution systems. *Water Sci. Technol. Water Supply* **2002**, *2*, 271–280. [CrossRef]
42.  Norton, C.D.; LeChevallier, M.W. Chloramination: Its effect on distribution system water quality. *Am. Water Works Assoc. J.* **1997**, *89*, 66–77. [CrossRef]
43.  Kumpel, E.; Nelson, K.L. Intermittent water supply: Prevalence, practice, and microbial water quality. *Environ. Sci. Technol.* **2015**, *50*, 542–553. [CrossRef]
44.  Xin, K.; Zhou, X.; Qian, H.; Yan, H.; Tao, T. Chlorine-age based booster chlorination optimization in water distribution network considering the uncertainty of residuals. *Water Supply* **2018**, *19*, 796–807. [CrossRef]
45.  Tryby, M.E.; Boccelli, D.L.; Uber, J.G.; Rossman, L.A. Facility location model for booster disinfection of water supply networks. *J. Water Resourc. Plan. Manag.* **2002**, *128*, 322–333. [CrossRef]

MDPI

St. Alban-Anlage 66

4052 Basel

Switzerland

Tel. +41 61 683 77 34

Fax +41 61 302 89 18

www.mdpi.com

*Water* Editorial Office

E-mail: water@mdpi.com

www.mdpi.com/journal/water

Lightning Source UK Ltd.
Milton Keynes UK
UKHW051812170822
407439UK00002B/109